Acid Rain

The Relationship between Sources and Receptors

Acid Rain

The Relationship between Sources and Receptors

Proceedings of a conference sponsored by:
Center for Environmental Information, Inc., Acid Rain Information Clearinghouse
33 South Washington Street, Rochester, New York 14608

Edited by
James C. White, Ph.D.
Professor Emeritus, Cornell University
Scientific Consultant to the Acid Rain Information Clearinghouse

Technical Editor
Carole N. Beal

Elsevier
New York • Amsterdam • London

Elsevier Science Publishing Co., Inc.
52 Vanderbilt Avenue, New York, New York 10017

Sole distributors outside the United States and Canada:

Elsevier Applied Science Publishers Ltd.
Crown House, Linton Road, Barking, Essex IG11 8JU, England

Library of Congress Cataloging-in-Publication Data

Acid rain.

1. Acid rain – Congresses. 2. Acid rain – Law and legislation – Congresses. I. White, James C. II. Acid Rain Information Clearinghouse (U.S.)
TD 196.A25A34 1988 363.7'3862 87-15595
ISBN 0-444-01277-X

Current printing (last digit)
10 9 8 7 6 5 4 3 2 1

Manufactured in the United States of America

Contents

Preface

This book is a compilation of papers presented at the third annual conference sponsored by the Acid Rain Information Clearinghouse (ARIC), held in Washington, D.C., December 3–4, 1986. ARIC is a project of the Center for Environmental Information in Rochester, New York.

We wish to extend our appreciation to the Electric Power Research Institute, the Gas Research Institute, and the National Acid Precipitation Assessment Program for their generous support in helping to underwrite the cost of the conference.

The conference was organized with the recognition that, if we are to have an environmentally sound and cost-effective control program, it is important to understand the relationship between sources and receptors of acidic deposition. In this era of fulminating information on all the aspects of acid deposition, it is extremely difficult for those working in any one area to appreciate and understand the opinion and problems of those in other areas.

This is especially true of the complicated source–receptor relationship, which is a problem of geography, meteorology, chemistry, physics, biology, law, and politics, to name only a few aspects. Most involved parties agree that emitted pollutants, especially sulfur and nitrogen, come down as acid deposition, but there is great disagreement as to where, when, and how they descend and what, if anything, they affect.

This book is designed for a nontechnical audience as well as specialists in acid rain research. The papers cover the current legal status and the nature and scope of scientific understanding and research programs, identify areas of consensus and disagreement, and assess policy options in the light of current understanding.

Because of the interdisciplinary nature of the topic, we asked Mary and Robert Pratt, editors of the *Acid Precipitation Digest,* to summarize the presented papers. Their short but accurate overview of the volume should encourage the reader to go deeply into the rest of the book.

It is extremely gratifying and indicative of the importance of the problem that the conference attracted twenty cosponsors representing very diverse interests, but we wish to state that the opinions expressed are those of the authors and do not necessarily reflect the views of the Center for Environmental Information or the conference cosponsors.

A most capable and distinguished program organizing committee, whose names are listed in these pages, did an exemplary job of identifying those people best able to clarify the picture. We wish to thank these authors for their cooperation, either in submitting camera-ready copy or in reading and editing the transcriptions taken from their taped presentations.

The editor is indebted to the Center for Environmental Research at Cornell University for providing office space and support services.

The staff of the Center for Environmental Information has performed with its usual efficiency. Linda Wall coordinated all conference arrangements. Lori Lopa patiently typed numerous drafts. William Wagner accomplished the multitudinous tasks necessary to assemble all the parts of this book, including charts, graphs, and textual material. Carole Beal was responsible for the technical editing, including draft revisions, proofreading, and assembly. Her painstaking attention to detail was essential to completing the editing in timely fashion. And lastly, without the enthusiasm, initiative, and organization of Elizabeth Thorndike, the conference and this volume would never have happened.

This book will probably not answer completely any question on acid deposition, but it does bring together the best thinking of the most capable people on all sides of the issues and points us toward some solutions.

James C. White

About the Center for Environmental Information

The Acid Rain Information Clearinghouse is a project of the Center for Environmental Information, Inc. (CEI), 33 South Washington Street, Rochester, New York 14608. CEI, a private, nonprofit organization established in 1974, provides information through its publications, access to computerized information sources, educational programs, conferences, and library. As a matter of policy, CEI does not take positions on environmental issues.

Conference Advisory Committee

Mohamed T. El-Ashry
World Resources Institute

Philip Galvin
New York State Department of Environmental Conservation

Maris Lusis
Ontario Ministry of the Environment

Lester Machta
National Oceanic and Atmospheric Administration

Michael Oppenheimer
Environmental Defense Fund, Inc.

Ralph M. Perhac
Electric Power Research Institute

Paul Ringold
National Acid Precipitation Assessment Program

Perry J. Samson
University of Michigan

James C. White
Acid Rain Information Clearinghouse

Conference Cosponsors

Cosponsors:

Air Pollution Control Association
American Meteorological Society
Cornell University Center for Environmental Research
Edison Electric Institute
Electric Power Research Institute
Environment Canada
Environmental Defense Fund, Inc.
Gas Research Institute
Motor Vehicle Manufacturers Association
National Acid Precipitation Assessment Program
National Association of Manufacturers
National Coal Association
National Council of the Paper Industry for Air and Stream Improvement
New York State Adirondack Park Agency
New York State Department of Environmental Conservation
Ontario Ministry of Environment
Quebec Ministry of Environment
Society of American Foresters
United States Environmental Protection Agency
World Resources Institute

Funding Support:

Electric Power Research Institute
Gas Research Institute
National Acid Precipitation Assessment Program

Acid Rain

The Relationship between Sources and Receptors

Acid Rain: The Relationship between Sources and Receptors
A Synopsis

Mary T. Pratt and Robert W. Pratt
R.D. 1 – Box 185
Valley Falls, NY 12185

Acidic deposition is one of the more scientifically complex and subtle environmental issues to emerge in the last decade. Although sulfuric and nitric acids formed in the atmosphere from sulfur and nitrogen oxides can affect health and damage natural resources and materials near their sources, the political debate over acidic deposition has differed from many environmental issues by centering around sources geographically far removed from sensitive receptors. If a program to control acidic deposition is implemented, the economic and social costs will be high; however, the long-term costs from the effects of acidic deposition could also be high.

Precisely because of these attributes, the issue has been debated vociferously over an extended period of time by those favoring or opposing a national program to reduce further the emissions of sulfur dioxide and nitrogen oxides -- the main precursor pollutants of atmospheric deposition of sulfuric and nitric acids. That acid deposition occurs and leads to certain adverse environmental effects is now an acknowledged fact by both sides in the debate, but agreement on the severity of the problem and its long-term effects has yet to be resolved.

One crucial component of the debate is reflected in the title of the conference -- "Acid Rain: The Relationship Between Sources and Receptors" -- organized by the Acid Rain Information Clearinghouse of Rochester, New York, and held in Washington, D.C., December 3 and 4, 1986. Intended for those with a general knowledge of the issue as well as specialists in acid rain research, the conference examined several thematic questions:

- Why do we need to understand source-receptor relationships?
- How do we learn about source-receptor relationships?
- What information about source-receptor relationships is still needed to explain the impacts of acidic deposition on aquatic and terrestrial ecosystems?
- How are source-receptor relationships used to devise control strategies?
- Given the incomplete nature of our understanding of source-receptor relationships, can the Environmental Protection Agency allocate emission reductions to control acid deposition through its authority under the Clean Air Act, or is new legislation needed?
- What questions has Congress been considering in drafting control legislation?

Acid Rain: The Relationship between Sources and Receptors
James C. White, Editor

A source-receptor relationship may be simply defined, as Ralph Perhac did: "Who does what to whom?" Or more technically: "What is the quantitative relationship between emissions and receptor exposure?"

Yet actually determining which emission sources of sulfur dioxide and nitrogen oxides are responsible for deposition of nitric and sulfuric acids at receptors hundreds of kilometers away is a difficult matter. By comparison, if a sewage treatment plant discharges pollutants into a surface water body such as a river, determination of source-receptor relationships there is generally much simpler than for acidic deposition in the atmos-phere. Rivers have fixed boundaries which confine both the transport of pollution and the receptors it can affect.

For the atmosphere, this is an extremely complex and uncertain undertaking, as several speakers noted. Because chemical reactions take place as the pollutants are being transported from emission sources, atmospheric chemistry and meteorology are linked in a complex manner that depends on the types of pollutants emitted as well as wind patterns, precipitation events and vertical transport of pollutants. The relationships vary from day to day, from season to season and even from year to year. To further complicate the analyses, the effects of acid deposition upon a receptor depend on the nature and location of the receptor itself.

Nevertheless, speakers felt it is important to determine source-receptor relationships. Charles Carter noted that the present Clean Air Act requires the source of air pollution producing a given effect to be determined before the Environmental Protection Agency may regulate the pollutant (although some other speakers disagreed). If Congress were to enact an acid deposition control program that would reduce acidic deposition in only selected, sensitive receptor areas, knowing source-receptor relationships would be essential for designing an effective emissions control strategy, Perhac stated. Even if a blanket emissions reduction policy were enacted, he said, knowing source-receptor relationships would, for example, help predict how much a receptor's exposure to acidic deposition would be reduced.

It is not surprising that the study of source-receptor relationships in acidic deposition is fraught with complexity and uncertainty -- a point scientists at this conference repeatedly made. Some speakers addressing policy and legal considerations took those themes and maintained that action can be, and in the case of Canada has been, taken despite uncertainty and complexity. Others began from the same point and reached opposite conclusions.

The major scientific tools used to determine source-receptor relationships are field experiments involving tracer releases, deposition monitoring, and the mathematical modeling of meteorological and chemical processes.

Deliberately releasing tracers into the atmosphere from known points and following their transport is one way to study source-receptor relationships, Lester Machta pointed out. Certain isotopes of sulfur might make ideal tracers, but they are radioactive and therefore socially unacceptable. During the Cross Appalachian Tracer Experiment perfluorocarbon compounds were released at Toledo, Ohio, and Sudbury, Ontario, and followed downwind. An elaborate tracer experiment to have involved the federal and the private sectors was proposed by the Electric Power Research Institute; to have been called MATEX, it was not implemented because of its high cost compared to its likelihood of success. Perry Samson mentioned that signatures based on characteristic ratios of the types of crustal elements found in different geographical areas have also been studied as naturally occurring tracers.

Machta detailed two other approaches. Historical data on the monitoring of emissions and deposition may be used. For example, sulfur dioxide emissions have declined annually over the last 15 to 20 years, and a decrease in sulfate concentration in precipitation has also been measured. One might think that a deliberate reduction in deposition would benefit sensitive receptors, except for an observation that introduces uncertainty into that conclusion: the concentrations of calcium and sodium in rainwater (components in precipitation unlikely to be related to human activity) have declined even more rapidly than has the sulfate concentration, leading to speculation on whether observed sulfate deposition decreases are the result of changes in emissions or some unexplained natural occurrence.

Sources of sulfur dioxide emissions could be deliberately reduced, Machta explained, and the changes in sulfuric acid deposition monitored at various receptor sites. High economic costs would be involved, and even if the number of emission sources to be modulated could be limited through statistical selection, the cost might still be too high for the experiment to be feasible.

The detailed and complex chemistry of the atmosphere as it relates to source-receptor relationships was discussed by Jeremy Hales. The information is important because reactions in the atmosphere affect the chemical composition, amount and geographic distribution of the deposited material. In order to produce detailed mathematical models of source-receptor relationships, the complex web of atmospheric chemical reactions must be understood.

The reactions of interest are those by which the pollutants sulfur dioxide and nitrogen oxides are oxidized into sulfuric and nitric acids. They cannot be treated separately, for the formation of sulfuric acid partly depends upon the presence of oxides of nitrogen. The formation of both acids is related to the presence of ozone and hydrocarbons, both of which derive partly from anthropogenic sources. Reactions in the atmosphere can take place in clear air (the gas phase), within water droplets (the aqueous phase), and even on the surface of ice particles in clouds.

For the gas phase, an overly-simplified sequence in the complicated chain of events might look like this. Light can trigger the dissociation of several atmospheric chemical compounds such as nitrogen dioxide, ozone, nitrous acid, or hydrogen peroxide to form a variety of chemical species including oxygen atoms or hydroxyl radicals (extremely reactive molecular fragments). Hydroxyl radicals can also be formed when oxygen atoms attack species like water, hydrocarbons or hydrogen peroxide. Although the mechanisms are complicated and not totally resolved, Hales said, sulfur dioxide, nitric oxide or nitrogen dioxide can be attacked by the hydroxyl radical to form sulfuric, nitrous and nitric acids. Thus atmospheric chemistry weds sulfur, nitrogen, carbon and other species.

Although gas-phase atmospheric chemistry is fairly well understood, much research remains to explain atmospheric aqueous-phase chemistry. Wet deposition of pollutants has been reliably measured for some time; however, the mechanisms and extent of dry deposition are still under investigation. Hales also described how atmospheric chemicals can react within storm systems, and how those storms can redistribute chemical species vertically and horizontally.

Because the details of atmospheric chemistry are too complicated to incorporate directly into computer models that also include transport, the research is used to tell which of the chemical processes are most important and which can be ignored. Simplifications are then made and that information helps to make the mathematical models of transport, transformation and deposition more efficient.

Considerable effort has been spent to develop mathematical models of long-range source-receptor relationships. The simplest of these, the Lagrangian models, Perry Samson explained, attempt to represent mostly meteorological transport and assume atmospheric chemistry behaves much more simply than it actually does; if they incorporate information about the ways pollutants react, they also assume the reactions are linear.

Because of small-scale atmospheric motions, a plume of emissions from a source is not transported neatly along a narrowly defined corridor, so modelers describe these pathways or trajectories as probability fields. The errors in the calculated trajectories can be considerable; for example, the estimated location for pollutants can be in error horizontally by 250 kilometers 24 hours after they have been emitted. So the farther removed receptors are from sources, the more uncertainty there is in determining source-receptor relationships.

While the Lagrangian models are used to predict source-receptor relationships over times as long as a year, Eulerian models like the Regional Acid Deposition Model (RADM) under development at the National Center for Atmospheric Research predict source-receptor relationships for very short periods, on the order of three days. Because they incorporate detailed chemical and weather information, they use large amounts of time on the largest computers and are expensive to run, so fewer numbers of cases can be investigated.

Samson feels that models like RADM will not give much new information about source-receptor relationships. Although knowing atmospheric chemistry is important, he said, knowing pollutant transport is the key for estimating source-receptor relationships over the long term. He noted that though complex, the influence of the weather on source-receptor relationships is not random. There is a "natural tendency to pollute." Because of the prevailing meteorology between sources and receptors, emission sources in certain regions are more likely to contribute deposition to receptors in other areas. For example, sources to the southwest of the Adirondack Mountains are more likely to contribute to acid deposition than are those in other sectors.

Despite the uncertainties in modeling and atmospheric chemistry, three types of emission reduction strategies have been developed. David Streets described them as emission optimized (emissions are reduced where they are greatest), cost optimized (emissions are reduced where least expensive to do so), and deposition-optimized (emissions are reduced cost-effectively to minimize the deposition to sensitive receptors). These strategies are increasingly more difficult to accomplish because the analyses are increasingly more difficult. However, the more complex analytical methods have the potential for achieving the most efficient control. Only the deposition-optimized strategy takes advantage of source-receptor relationships.

Both Streets and James Young presented the results of calculations showing how the costs of emission reductions vary with the degree of reduction of acids deposited on sensitive receptors. In the case of the Adirondack Mountains, Streets said, the cost of achieving a target loading of 20 kg/ha/yr is in the neighborhood of $2 billion per year. Further decreases in deposition come at much higher costs. The cost of reducing the target loading to 15 kg/ha/yr is roughly double the cost for the 20 kg level.

Although all but one of the acid rain control proposals introduced by Congress through 1986 have been emission optimized, those strategies actually put in place in North America so far (by Canada and several states) have been deposition optimized. They use transfer matrices coupled with computer models to decide which emission sources are responsible for acidic deposition to sensitive receptor areas, and how much these sources must cut back to reduce deposition to desired levels in the most cost-effective fashion.

To be fully exploited, deposition-optimized strategies rely on the establishment of receptor vulnerabilities to different pollutants. Considerable research has taken place to determine which ecosystems in North America are most vulnerable to acidic deposition, and which adverse effects have already taken place. As with the atmospheric component of the source-receptor relationship, the studies of terrestrial and aquatic ecosystems are very complex.

Gene Likens noted that it is more important to know how rapidly an aquatic ecosystem will be acidified, than the fact that it can undergo acidification; the overall amount of acids deposited is more important than their concentration in rainwater. The composition of acid deposition may also affect the degree of damage done. For instance, Likens and Laurence Kulp pointed out, aquatic and terrestrial ecosystems can utilize the nitrogen falling in acidic deposition in summer, but in winter it leaves the ecosystem. Likens concluded that reductions in deposition to as low as 10 kg/ha/yr (one half the level recommended by the U.S.-Canada Memorandum of Intent on Transboundary Air Pollution) may be necessary to protect some sensitive aquatic ecosystems.

Kulp showed how topography can influence the source-receptor relationship for a terrestrial ecosystem; for example, alpine forests receive acidic deposition from fogs and mists that is more acidic than rain falling at lower altitudes. Nevertheless, the role of these pollutants in observed forest declines has not been explained. Seasonal variations in atmospheric chemicals can also affect receptors; the strong oxidizing agent hydrogen peroxide is present in great quantities in summer in high altitude forests.

Episodes of elevated pollutant levels can also affect receptors more than if the same amount of pollutant were delivered over a longer period of time, Likens said. Pulses of high acidity are well documented in surface waters as the snow melts; the responses of organisms to these pulses will depend upon their varying sensitivities. In urban areas, episodes of high ozone concentrations are injurious to vegetation, Kulp explained, but it is not known if episodic elevations can occur in rural areas.

Depending upon the properties of an ecosystem, the effects of acid deposition may be delayed. Acidic compounds may accumulate, for example, until a soil's carrying capacity is exceeded. In the Southeast, Likens said, soil sulfur adsorption capacities are greater than in the Northeast, but sulfate release from these soils has recently been observed because their capacities to adsorb sulfur has been exceeded.

In his analysis of the scientific information presented at the conference, George Hidy concluded that although the ability to calculate source-receptor relationships has improved, the capability for testing and verifying calculations has not. He foresees little prospect for new data acquisition and dissemination before the 1990s. The National Acid Precipitation Assessment Program has lagged in publication of documents, he noted, and Congressional inaction has been reinforced by hiding behind the ambiguities of scientific information.

If Congress does not decide the matter of acid deposition, a resolution could come through the courts. Several northeastern states have sued to force the Environmental Protection Agency to take action under the present Clean Air Act. This conference reviewed those cases, the ways pollutant regulation is accomplished through the Clean Air Act, the Environmental Protection Agency's position on the issue, and the question of scientific uncertainty in legislative, regulatory and judicial processes.

David Wooley said it is important to recognize that the scientific debate on acid deposition takes place within a legal framework and that scientific uncertainty and regulatory action are not incompatible. The EPA is using "uncertainty in the science of source-receptor relationships. . .to resist demands for action and to obscure [its] obligations under existing law," he pointed out. In writing the 1977 amendments to the Clean Air Act, Congress concluded that imprecise regulation is preferable to no regulation. Past court actions have supported regulatory action despite uncertainty; the courts have been satisfied as long as the data and methods used were the best available at the time.

Michael Teague cited other examples from the courts to support his contention that uncertainties in source-receptor relationships make it impossible for the EPA to allocate sulfur dioxide emission reductions. For example, petitions that argue the sulfur dioxide emissions from one state adversely affect sulfate levels in another state, to a degree that prevents states from attaining the total suspended particulate standard, have not been successful. In examining several provisions of the Clean Air Act, he concluded that EPA does not now have the information needed, such as an approved mathematical model for sulfate control, to establish an ambient standard for sulfate because "unambiguous attribution of ambient sulfates to a specific source is impossible."

The Clean Air Act was not written like the technology-based Clean Water Act, which requires the best available technology to be used in controlling pollutant discharges. If it had been so, we might not be arguing about source-receptor relationships in acidic deposition, according to Charles Carter. He considers the Clean Air Act cumbersome to administer because the EPA is required to "concentrate on a particular effect or set of effects and establish a standard with an adequate margin of safety." He also said that the Clean Air Act as written in 1970 and amended in 1977 was crafted originally to gauge localized pollution based on ambient levels of pollutants. He also cited several legal and regulatory examples to support the view that the key to regulation under the Act is relating damage to existing sources.

In Congress, of the more than 40 acid deposition control proposals introduced through 1986, few have attempted to take advantage of source-receptor relationships -- rather they have been source-oriented measures, John Gibbons explained. Congressional staff seem to view source-receptor models as both a hindrance and an asset. One bill, introduced by Senator Christopher Dodd, would have prohibited the EPA from relying solely on models for regulating interstate transport of pollutants. A proposal by Senator David Durenberger, which failed to attract cosponsors, would have allocated emission reductions following technical analyses that included source-receptor models. However, Ralph Perhac pointed out that knowledge of source-receptor relationships will be useful even if a source-oriented strategy were eventually adopted by Congress, because we will still want to know what deposition reductions to expect from mandated emission reductions.

If Congress does act to control acidic deposition, Gibbons said, it must address several key questions, such as which pollutants to reduce, by what degree and what time scales and geographic regions. So far Congress has concentrated on social and economic questions relating to costs of controls and effects on employment.

Although this conference was intended to educate and inform, the question of the need for a control program inevitably arose. Michael Oppenheimer said that we do not require any more scientific information to begin controlling acidic deposition, but more research will guide a control program as it proceeds.

Laurence Kulp, in answer to a question, said that the emerging integrated coal gasification combined cycle combustion technology, which controls sulfur dioxide and nitrogen oxides better than scrubbers, will become the new coal combustion technology no matter what government does about acidic deposition. Now we must decide whether the effects of acidic deposition are such that we need a crash retrofit program or a more rapid replacement of combustion technology.

If the decision is made to proceed with emission controls, the public would be best served by legislation forcing new combustion technology, George Hidy said, rather than relying on retrofitting. Scientific opinion on acid deposition remains polarized, he noted, and calls for action still seem weakly justified by information on the effects of acidic deposition, compared with more pressing environmental issues.

Gene Likens' answer to a question following his presentation typifies the diversity of opinion expressed at this conference. Added information about the complexities of aquatic ecosystems is not necessary to take regulatory action, he said, acknowledging that scientific input contributes only 15 to 20 per cent of the decision making process.

SOURCE–RECEPTOR RELATIONSHIPS: LEGAL ASPECTS

Acid Rain: The Emerging Legal Framework

Neil Orloff* and Lisa Ann Byrns**
Center for Environmental Research
Cornell University, Hollister Hall
Ithaca, NY 14853

Congress is still in the early stages of developing legislation to control acid rain. During 1986, five bills[1] were introduced in Congress and hearings were held on them; but, no legislation reached the floor of either chamber. This pattern has been repeated since the early eighties. Each year legislation has been introduced that specifically addresses acid rain but, with one exception, no statute has yet been enacted.

In 1980, Congress passed the Acid Precipitation Act of 1980,[2] which created a task force to study the causes and effects of acid rain. This twenty-member group, known as the Acid Precipitation Task Force, is required to develop and carry out a 10-year comprehensive research plan under a total appropriation of $50 million. This research plan includes a nationwide, long-term monitoring network measuring levels of acid rain and a program studying emissions sources which contribute to acid precipitation. The plan also addresses both long-range atmospheric transport models and economic assessments. These assessments concern the environmental impact of acid precipitation and the alternative technologies to remedy or ameliorate acid rain's harmful effects. The Act does not impose any limitations on emissions.

Thus, as 1987 unfolds, Congress is still in the process of developing a foundation for the eventual passage of control legislation. At the present time, there is no consensus on the specific provisions of this legislation.

---

*Neil Orloff is the Director of the Center for Environmental Research at Cornell University and Of Counsel to the Los Angeles law firm of Irell & Manella.

**Lisa Ann Byrns is currently a third-year law student at Cornell University.

[1] H.R. 4567, 99th Cong., 2d Sess. (1986); S. 2003, 99th Cong., 2d Sess. (1986); S. 2200, 99th Cong., 2d Sess. (1986); 99th Cong., 2d Sess. (1986); S. 2203, 99th Cong., 2d Sess. (1986); S. 2813, 99th Cong., 2d Sess. (1986). Representative Charles Whitley (D-NC) introduced H.R. 2631 in 1985 and the House of Representatives approved the bill by a vote of 416 to 4 on August 13, 1986. The measure was then sent to the Senate Agriculture Committee, which took no action on it. H.R. 2631 would have established a ten-year program to study the effects of acid rain and air pollution generally on U.S. forests.

[2] 42 U.S.C. §§ 8901-8912 (1982).

Acid Rain: The Relationship between Sources and Receptors
James C. White, Editor

In the absence of this consensus, efforts at the federal level to control acid rain have rested on existing legislation -- the Clean Air Act.[3] While not specifically designed to address the acid rain problem, the Act does contain provisions that are susceptible of being employed to control the long-range transport of pollutants.

This paper briefly describes: (1) the efforts in Congress during 1986 to develop new legislation to control acid rain, (2) the provisions of the Clean Air Act that states and environmental groups have used as an interim solution pending the enactment of new legislation by Congress, and (3) efforts at the state level to address the problem.

ACTIVITY IN THE HOUSE OF REPRESENTATIVES

During this past year, the dominant activity in Congress centered in the House of Representatives. Attention focused on H.R. 4567[4] -- The Acid Deposition Control Act of 1986. Introduced by Congressman Henry Waxman (D-Calif) and co-sponsored by a bipartisan coalition of more than 150 other members of the House, the bill would require, by 1997, reductions in sulfur dioxide emissions of 10 million tons per year and in nitrogen oxide emissions of 4 million tons per year. The bill would achieve these reductions by imposing stringent emission limitations on boilers in power plants, boilers in industrial facilities, and on motor vehicle emissions. Estimates of the cost of complying with these emissions reductions range from $3 billion per year to $10 billion per year.

H.R. 4567 separates the emissions reductions into two phases. The first, ending in 1992, would require a 5-million ton reduction of sulfur dioxide emissions and a 2-million ton reduction of nitrogen oxide emissions. The bill requires the EPA to study and report to Congress in 1993 on the reductions achieved during the first phase and on the feasibility of meeting the remaining requirements by 1997, the deadline for the second phase.

The emission reductions would be partially financed by a tax on electric utilities. H.R. 4567 authorizes the EPA Administrator to impose a fee not to exceed 1/2 mill per kilowatt hour on the "generation and importation of electric energy"[5] except from hydroelectric and nuclear power plants. This fee can be imposed only from December 31, 1988 to December 31, 1996. The bill gives EPA substantial enforcement authority,

[3] 42 U.S.C. §§ 7401-7626 (1982).

[4] H.R. 4567, 99th Cong., 2d Sess. (1986).

[5] Id. at § 185(a).

allowing the agency to impose a $50,000 per-day civil penalty or institute a civil action for nonpayment of the fee.

Collected fees would be deposited in the Acid Deposition Control Fund, which would serve as an interest subsidy program for electric utilities administered by EPA. H.R. 4567 recognizes that utilities would most likely incur debt by installing controls and other technologies to comply with the bill's reduction requirements. The Fund's proceeds would be used to pay a portion or all of the utility's interest on this "qualified pollution control debt."[6] The subsidy is to protect the electric utility's residential customers from excessive rate increases, which H.R. 4567 defines as increases exceeding 10% of the rate which would have been billed if no qualified pollution control debt had been incurred.

The Subcommittee on Health and Environment held hearings on the bill throughout April and May of last year. Supporters of the measure included Governor Earl of Wisconsin, the State and Territorial Air Pollution Program Administrators, and representatives of environmental organizations.

In contrast, industry groups voiced substantial opposition to the bill. Coal, electric, aluminum and steel industry representatives testified that the proposed legislation carried high costs and negative competitive effects. The United Mineworkers of America also opposed the bill. Auto industry spokesmen asserted that the legislation would increase the price of cars and trucks by $100 to $250 per vehicle. The Edison Electric Institute, a lobbying group of electric utilities, claimed the bill would cost electric utilities $9.2 billion annually and that eligibility for the interest subsidy program would be minimal. The Institute predicted that utility rates would exceed 10%, but these increases would not qualify for the subsidy program. The subsidy would be available under H.R. 4567 only when the utility's increased rates result from incurring and servicing pollution control debt. The group's study asserted that debt repayment alone would not increase rates by 10%, although other costs -- such as fuel-switching -- would precipitate rate hikes exceeding H.R. 4567's threshold level of 10%. On April 29, 1986, the Congressional Research Service released a report that concluded that the group's $9.2 billion estimate failed to account for the more cost-effective compliance strategies that were available. This report stressed the bill's flexibility which allowed states to design their own strategies for emissions reductions.

Federal government officials opposed H.R. 4567. EPA Administrator Thomas stated that an inadequate scientific foundation demonstrating acid rain damage formed the

---

[6] Id. at § 187(a), (c).

basis for the bill's emissions reductions. Energy Secretary Herrington criticized H.R. 4567. West Virginia Governor Moore also expressed opposition, testifying that the measure would result in a loss of mining jobs and a decrease in annual coal production in his state.

On May 20, 1986, the Subcommittee on Health and the Environment approved the bill by a 16 to 9 vote. The bill was reported to the full Energy and Commerce Committee; but, the full committee never reported the bill to the House floor.

In approving the measure, the subcommittee accepted several amendments to H.R. 4567. The most important of these concerned "emissions averaging" and the interest subsidy program. Sulfur dioxide emissions would continue to be averaged on an annual basis rather than the monthly basis originally required in the bill. Further, the subsidy payments would no longer be limited to servicing the electric utility's qualified pollution control debt. The amendment authorized EPA to disperse funds to cover rate increases resulting from either interest payments or other costs incurred in complying with the emissions reductions. Thus, for any increased costs due to compliance with H.R. 4567, electric utilities would be subsidized to limit subsequent residential rate increases to 10%. A state governor would have to establish two criteria to EPA's satisfaction before funds would be dispersed. First, that the residential rate increases attributable to compliance costs are "substantially equivalent" throughout the state. Second, that these increases were "substantially levelized" among the state's utilities while the reduction requirements were in effect.

The "levelizing amendment" prompted the House Subcommittee on Energy, Conservation and Power, also a panel of the Energy and Commerce Committee, to hold hearings in June on H.R. 4567. Electric utility representatives and EPA Administrator Thomas testified, once again voicing opposition to the bill. Federal Energy Regulatory Commission Acting Chairman Sousa stated that H.R. 4567 would involve FERC, EPA, state governors, and public utility commissions in rate-setting activities, which might be unworkable. Public utility commission officials from Indiana and West Virginia testified that state law could prohibit a governor from engaging in the rate-making activities required under H.R. 4567. On August 14, 1986, the Subcommittee sent the bill back to the full Committee without any changes.

While the bill was still technically under the jurisdiction of the Energy, Conservation and Power Subcommittee, the full Committee began markup of H.R. 4567 on August 13, but it adjourned the following day. No further action was taken on H.R. 4567 before the close of the 99th Congress.

ACTIVITY IN THE SENATE

In the Senate, two bills were the focus of activity -- S. 2203[7] and S. 2813.[8]

S. 2203 -- the more stringent bill -- would require reductions in sulfur dioxide emissions of 12.3 million tons per year. Unlike legislation offered in the previous Congress which applied only to 31 eastern states, this bill would impose nationwide emissions reductions. S. 2203 also proposed stronger controls on hydrocarbon and carbon monoxide emissions to prevent forest damage. S. 2203 was introduced by Senator Stafford (R-Vt) on March 18, 1986.[9]

S. 2813 was patterned after H.R. 4567's two-phase reduction levels and would require reductions in sulfur oxide emissions of 10 million tons per year. Unlike H.R. 4567, however, S. 2813 would not impose a nationwide electricity tax nor offer utilities a subsidy for increased costs. Instead, the full costs of meeting emission limits would be placed on the utility. S. 2813 was introduced by Senator Proxmire (D-Wis.) in August, 1986.

The Senate Committee on Environment and Public Works held hearings on these two bills in September and October. It did not, though, report either of these bills to the floor to the Senate.

EXISTING LEGISLATION: THE CLEAN AIR ACT

In the absence of new legislation directly addressing the problem of acid rain, efforts to abate acid rain have focused on existing legislation -- the Clean Air Act.[10] The framework of the Clean Air Act was established in 1970 and revised in 1977. It views air pollution primarily as a problem of reducing ambient levels of pollutants through imposing

[7]S. 2203, 99th Cong., 2d Sess. (1986).

[8]S. 2813, 99th Cong., 2d Sess. (1986).

[9]On the same day, Senator George Mitchell (D-Maine) introduced S. 2200, which would have established a long-range transport corridor consisting of the states east of the Mississippi River. The bill proposed to limit sulfur dioxide and nitrogen oxide emissions in this corridor to 1981 levels. S. 2200 was sent to the committee on Environment and Public Works. Other acid rain legislation, S. 2003, had been introduced in January, 1986, by Senator Daniel Moynihan (D-NY), which was referred to the same Committee. The Committee took no action on either S. 2003 or S. 2203.

[10]42 U.S.C. §§ 7401-7626 (1982).

emissions limitations on sources in the immediate vicinity of these high pollution concentrations. The framework was not designed with the long-range transport of pollutants in mind.

Under the Clean Air Act, EPA is directed to promulgate national ambient air quality standards (NAAQS), identifying the maximum concentration of specific pollutants consistent with protecting public health and welfare. Each state is then required to develop a state air implementation plan (SIP), which imposes emissions restrictions on sources within the state so that the concentration of pollutants in the state are brought down to the level of the ambient standards. Thus, the air pollution program heavily depends on linking "sources" and "receptors" within the immediate vicinity of each other. With only a few exceptions, no emission requirements are imposed on sources within a state except to the extent necessary to produce a corresponding reduction in ambient pollution concentrations within the state.

Two exceptions to this general theme of the Clean Air Act form the foundation of current efforts to control acid rain.

International Air Pollution

Section 115[11] of the Act addresses international air pollution. It requires a state to control in-state polluters whose emissions cause or contribute to air pollution which "may reasonably be anticipated to endanger the public health or welfare of a foreign country." Whenever EPA's Administrator, on the basis of either a report by a "duly constituted international agency" or a request by the Secretary of State, has reason to believe that pollutants emitted in the United States may endanger a foreign country, the Administrator must formally notify the governor of the state from which the emissions originate. This, in turn, triggers a state's obligation to revise its implementation plan. Upon receipt of the formal notice, the state must revise as much of its SIP as is necessary to prevent or eliminate the air pollution endangerment to the foreign country.

These seemingly straightforward procedures were contested in the case of Thomas v. New York,[12] decided on September 18, 1986 by the U.S. Court of Appeals for the District of Columbia Circuit. The case arose from January 13, 1981 letters written by outgoing EPA Administrator Douglas M. Costle to then Secretary of State Edmund Muskie

[11] 42 U.S.C. § 7415 (1982).

[12] 802 F.2d 1443 (D.C. Cir. 1986).

and Senator George Mitchell (D-Maine). Administrator Costle expressed his belief that pollution emitted in the United States was at least partially responsible for acid rain endangering the public welfare of Canada. Mr. Costle based his finding of endangerment on a report issued by the International Joint Commission, an international agency.

No action was taken by Costle's successor -- Anne Gorsuch. A Clean Air Act "citizen's suit" was then instituted by seven eastern states, four national environmental groups, a Congressman, and two American citizens who owned property in Canada. Section 304[13] of the Act authorizes any person to institute a civil action against the Administrator for failure to perform a nondiscretionary act or duty required by the statute. The plaintiffs argued that the Costle letters imposed upon the successor EPA Administrator a duty to identify the states responsible for acid deposition and to issue SIP revision notices to these states. The District Court agreed.[14] However, the Court of Appeals reversed in an unanimous opinion written by now Supreme Court Associate Justice Antonin Scalia.[15]

The Court of Appeals held that an EPA determination that air pollutants affect the public health or welfare of a foreign country is an instance of rule-making. This requires that EPA follow procedures of notice, comment, and Federal Register publication. Because Mr. Costle sent the letters without allowing for advance notice and publication, nor providing a comment period, the correspondence would not operate to the trigger mandatory agency required by Section 115.

The significance of this decision lies in its expansive reading of Section 115. The Court of Appeals broadly applies formal procedural requirements to EPA's decision-making under Section 115. The statutory language on its face requires the Administrator to "have reason to believe" air pollutants from the U.S. endanger a foreign country's public health or welfare. The court interprets this statutory language as requiring that rule-making procedures be used to find endangerment. Based upon an international agency report, an Administrator may believe that endangerment exists. Mere belief, according to the court, will not trigger SIP revision notices. Rather, the court's holding requires EPA to undertake a rule-making process to find endangerment before issuing formal notification to the state. In effect, this will result in one instance of rule-making prior to a second instance of related rule-making.

---

[13]42 U.S.C. § 7604(a)(2) (1982).

[14]State of New York v. Thomas, 613 F. Supp. 1472 (D.D.C. 1985).

[15]Thomas v. State of New York, 802 F.2d 1443 (D.C. Cir. 1986).

Interstate Air Pollution

While the Clean Air Act focuses primarily on reducing local emissions in order to reduce local ambient air concentrations, the statute does address interstate air pollution. Certain provisions seek to guarantee that air pollution generated in one state does not disrupt another state's plans for complying with the NAAQS. The Act permits a state to challenge emissions restrictions imposed on sources in other states when those restrictions would interfere with the challenging state's attainment of air quality standards. Sections 110[16] and 126[17] speak to this problem of interstate air pollution.

Under Section 110, EPA must evaluate each state implementation plan on the basis of eleven criteria. These criteria ensure that the plan contains specific provisions, such as a permissible time-frame for attaining the state's NAAQS, schedules for emissions limitations, plan revision procedures, enforcement programs, and adequate personnel and funding to carry out the SIP. One criterion requires that the plan contain adequate provisions prohibiting any new or existing in-state stationary source from emitting air pollutants in amounts which would prevent another state from maintaining or attaining the state's NAAQS.

A state must obtain EPA approval each time a plan or revision is proposed, e.g., to change stack heights or in-state emissions levels. Agency approval involves formal notice, comment and Federal Register publication, which affords another state the opportunity to comment on the proposed plan or revision. The commenting state may argue that the proposal fails to meet the Section 110 criterion concerning interstate-effects of the SIP and that the commenting state will be unable to maintain or attain its NAAQS if the proposal is approved.

Section 126 offers a state another avenue to challenge an implementation plan. Any SIP may be challenged, not only those plans or revisions seeking EPA approval. A state or municipality may petition EPA to find that emissions in another state from a new or existing source will interfere with the petitioning state's maintenance or attainment of air quality standards. If EPA makes such a finding, Section 126 prohibits construction of the new source (even if state permit has been issued) or continued operation of the existing source -- subject only to "phase-in" provisions. For example, the Act grants EPA authority to permit

---

[16]42 U.S.C. § 7410 (1982).

[17]42 U.S.C. § 7426 (1982).

the continued operation of an existing source for a limited period if the source complies with additional emission limitations and compliance schedules that EPA may impose on it.

In this context, the States of Connecticut and New York and Jefferson County, Kentucky, brought separate suits against EPA in federal courts challenging agency plan approval under Section 110 or seeking judicial review of EPA decisions on Section 126 petitions. Each petitioner argued that the SIP at issue would interfere with the maintenance or attainment of ambient air quality standards in the petitioner's jurisdiction. Although each was unsuccessful in its Section 110 or Section 126 assertion, each court reached its result by a different route.

The State of New York brought suits against EPA challenging the agency's Section 110 approval of air implementation plans for the States of Illinois[18] and Tennessee.[19] EPA's approvals permit increases in sulfur dioxide emissions from utilities in each of those states. The New York challenges argued that emissions increases from these states interfere with maintenance or attainment of ambient air quality standards in New York.

In each suit, New York contended that when EPA reviews a SIP revision, the agency must consider the cumulative impact of all emissions from the state -- even though the SIP revision may apply to a single source. Both courts rejected this contention. The Courts of Appeals for the Sixth and Seventh Circuits based their results on their views of the agency's discretion. The courts ruled that, as long as the narrower criteria for approval set forth in Section 110 are met, the agency has discretion to limit the scope of inquiry into the proposed revision to the specific emission changes giving rise to the revision.[20] Each court also noted that New York's pending Section 126 petitions, not suits under Section 110, were the proper forum to determine the cumulative impact of interstate pollution.

The State of New York has also had its own SIP revision disputed in court under Section 110. The State of Connecticut challenged EPA's approval of New York's plan revision, which permitted the continued use of 2.8% sulfur content fuel at five Long Island power plants.[21] New York's general environmental scheme authorized burning 1% sulfur

---

[18] State of New York v. U.S. Envtl. Protection Agency, 716 F.2d 440 (7th Cir. 1983).

[19] State of New York v. U.S. Envtl. Protection Agency, 710 F.2d 1200 (6th Cir. 1983).

[20] 716 F.2d at 442; 710 F.2d at 1204.

[21] State of Connecticut v. U.S. Envtl. Protection Agency, 696 F.2d 147 (2d Cir. 1982).

content fuel in this area. Connecticut argued that it would be harmed by the New York revision and would be unable to maintain or attain its NAAQS. Connecticut specifically challenged EPA's statistical modeling, technical data, and meterological calculations. The Second Circuit Court of Appeals rejected these arguments and found that EPA neither abused its discretion nor violated the Act in approving New York's revised SIP.

The Second Circuit refused to require EPA, in the process of reviewing a proposed revision, to make a cumulative impact study of all pollution sources. However, the court did require that EPA consider the effects of a specific SIP revision on attainment of ambient air quality standards in other states.[22] The court reasoned that the Section 110 criterion (concerning interstate effects of emissions) was not intended to prevent minimal impacts upon another state's pollution concentrations simply because that state had not attained the national standards. Therefore, the Second Circuit's approach can be fairly described as review to determine "minimal impact": a SIP revision impacting on a nearby state could be approved only where the impact is "so insignificant as to be... minimal."[23]

In another court challenge, Jefferson County, Kentucky, sought judicial review of EPA's order denying the county's Section 126 petition.[24] Jefferson County petitioned the agency to find that the SIP of neighboring Floyd County, Indiana permitted emissions which prevented Jefferson's attaining or maintaining its NAAQS. The court rejected the minimal impact review used by the Second Circuit and, instead, developed its own standard: "significantly contributes" to the petitioning state's inability to meet its NAAQS.[25] The agency had determined that emissions in Indiana contributed approximately three percent of the pollutants in Jefferson County that violate NAAQS. The court found that this impact could not significantly contribute to prevent the petitioning county from attaining or maintaining its NAAQS.

In the absence of new legislation to address acid rain, most of the legal battles have surrounded these two sections of the Clean Air Act. These battles have been unsuccessful for several reasons. First, neither the Act nor its legislative history offers guidelines on how to apply the statutory language "prevent the attainment or maintenance" of NAAQS.

---

[22] Id. at 163.

[23] Id. at 165.

[24] Air Pollution Control District v. U.S. Envtl. Protection Agency, 739 F.2d 1071 (6th Cir. 1984).

[25] Id. at 1093.

One albeit extreme interpretation of this language would count only discharges from an emitting state, assuming that the receiving state was making no contribution to the ambient level of pollution in the state. Another interpretation would also count the receiving state's own emissions in determining if there is interference with attaining or maintaining an air quality standard. Yet, this approach could require an upwind state to prohibit all emissions of a given pollutant where the receiving state had made maximum use of its permitted emission level for that pollutant.[26] The upwind state would be prohibited from contributing any amount of pollution to the receiving state's ambient concentrations.

Second, judicial review of agency action is narrow.[27] Courts must inquire if the action is arbitrary and capricious, constitutes an abuse of agency discretion, or is otherwise not in accordance with the law. But, great deference is given to the agency's interpretation of the statutes which it is administering.[28] Decisions under the Clean Air Act often involve masses of technical, mathematical, and statistical information, and complex models developed by EPA. With judicial deference to the agency, a petitioning state carries a heavy burden in court.

Finally, the Act deals primarily with pollution within a state, not cumulative emissions from a group of states. Although the Act develops air quality control "regions" and directs EPA to consider the interstate impact of a state implementation plan, a challenging state may only petition against another individual state. There is no concept under the Clean Air Act such as a regional implementation plan. While the Act's stated policy is to protect the nation against all harm from air pollution, there is no provision in the statute dealing specifically and primarily with long-range regional transport of air pollutants.

ACTIVITY AT THE STATE LEVEL

While most of the attention has been focused at the federal level, several states -- not content with waiting for the federal government to act -- have moved ahead in passing legislation that addresses the problem of acid rain. So far, seven states have enacted

---

[26]Id. at 1090 (citing Hirsch & Abramovich, Clearing the Air: Some Legal Aspects of Interstate Air Pollution Problems, 18 Duq. L. Rev. 53, 67 (1979)).

[27]State of Connecticut, 696 F.2d at 155.

[28]Id.

statutes that address the problem -- California,[29] Maine,[30] Minnesota,[31] New York,[32] New Hampshire,[33] Washington,[34] and Wisconsin.[35] These statutes adopt one or both of two approaches -- (i) they establish an acid rain research program, and/or (ii) they impose emission requirements on sources within the state beyond those required to attain federal ambient air quality standards.

The California Acid Deposition Act establishes a scientific advisory committee to assist the state's Air Resources Board in designing and implementing a comprehensive research and monitoring program. California's program is funded by both the state's Motor Vehicle Account and fees imposed on nonvehicular sources of sulfur dioxide and nitrogen oxide which emit more than 1,000 tons per year. The program's statutory goals are similar to, but more extensive than, those under the federal Acid Precipitation Control Act of 1980. Like that U.S. statute, the California legislation does not impose any limitations on emissions. Moreover, the California Act remains in effect only until December, 1988, unless further legislation is enacted.

The State of Maine has enacted legislation which authorizes a study of acid deposition but, like the California statute, does not establish emissions limitations. Maine requires its Department of Environmental Protection to inventory in-state emissions sources and to conduct an acid rain impact study. The Washington Clean Air Act also requires a comprehensive evaluation of acid rain within the state and directs the Department of Ecology to maintain periodic monitoring of acid deposition "to ensure early detection."

The State of Minnesota enacted legislation in 1982 designed "to mitigate or eliminate the acid deposition problem by curbing sources of acid deposition." The state's Pollution Control Agency obtained approval in 1986 for a two-phase reduction plan. In 1990, utilities in Minnesota meeting certain threshold requirements would be required to hold

---

[29]Cal. Health & Safety §§ 39900-39915 (West Supp. 1986).

[30]Me. Rev. Stat. Ann. tit. 38, §§ 603-B (West Supp. 1986).

[31]9 Minn. Stat. Ann. §§ 116.42-116.45 (West Supp. 1987).

[32]N.Y. Envtl. Conserv. Law §§ 19-0901 to 19-0923 (McKinney Supp. 1987).

[33]N.H. Rev. Stat. Ann. § 125-D:1 to D:3 (West Supp. 1986).

[34]Wash. Rev. Code Ann. §70.94.800 to 70.94.825 (West Supp. 1987).

[35]Wisc. Stat. Ann. §§ 144.375 to 144.389 (West Supp. 1987).

sulfur dioxide emissions to 130% of the utility's 1984 emissions levels. Further reductions would come in 1994.

The New York Acid Deposition Control Act also establishes a two-tiered schedule for emissions reductions. The Department of Environmental Conservation must establish interim control targets that sources of sulfur dioxide and nitrogen oxide emissions must meet by 1988. DEC must promulgate final control targets for sulfur dioxide emissions and nitrogen oxide emissions by 1991 and 1987, respectively. The Act, however, does not specify when these final control targets are to take effect. The statute also eliminates the scheduled requirements if a federal statute or program is enacted that DEC determines is consistent with the state legislation's purpose. In 1985, the State of New Hampshire enacted legislation similar to New York's. This statute requires a 25% sulfur dioxide reduction by 1990 and, if Congress enacts national acid rain control requirements, another 25% reduction by 1995.

The State of Wisconsin enacted legislation in April, 1986, to reduce emissions. Sulfur dioxide and nitrogen oxide emissions are subject to different time-tables and utilities will be required to submit annual compliance plans. After 1992, sulfur dioxide emissions from major utilities will be limited to 325,000 tons annually. In 1991, nitrogen oxide emissions from major utilities will be held at 135,000 tons annually.

Other states, while not enacting specific legislation, also addressed the issue of acid rain in 1986. The Colorado Acid Deposition Task Force reported to Governor Lamm in March, calling for additional research to substantiate present concerns that acid deposition affected the state's ecosystems. In June, New England states' governors approved an acid rain plan to reduce sulfur dioxide emissions in the region by 1995. The plan was also adopted by the premiers of the Eastern Canadian provinces. The Southern Governors Association unanimously adopted a resolution in August to support further acid rain research.

A draft uniform law, the Uniform Transboundary Pollution Reciprocal Access Act,[36] was approved by both the National Conference of Commissioners on Uniform State Laws and the Uniform Law Conference of Canada. This Act is jurisdictional in nature and does not create any new substantive rights or bases for relief. It provides access to the courts in one jurisdiction for pollution victims from another jurisdiction. This access is conditioned on reciprocity: both the polluting and the victim's jurisdictions must have enacted the Uniform Act or provided substantially equivalent court access. The Act is

[36]9A U.L.A. 465 (Supp. 1986).

intended to alleviate the hardship a pollution victim faces in gaining access to courts in the polluter's jurisdiction. Under the traditional "local action rule," an action for injury concerning land can only be brought in the jurisdiction where the land is located. Land may have been damaged by air pollution, but the polluter may not be a resident of the same jurisdiction where the land is located. The local action rule precludes the land-owner from filing a suit in the defendant's home state, yet other jurisdictional rules prevent an injured land-owner from hauling the out-state defendant into court in the land-owner's state, especially if the defendant is a foreign citizen. The Act ameliorates the land-owner's procedural dilemma by offering the land-owner the opportunity to file a suit in the state where the polluter, whether an individual or a corporation, resides. So far, three states have adopted this uniform law: Colorado,[37] New Jersey,[38] and Wisconsin.[39]

FUTURE DIRECTIONS

As the nation enters 1987, this patchwork quilt of weak provisions of the Clean Air Act and scattered state statutes constitutes the legal framework surrounding the control of acid rain. While consensus is often difficult to achieve in the field of environmental law, it currently exists among opposing parties on at least one aspect of this framework: it is woefully inadequate to address the problem of acid rain.

This consensus breaks down at the next step: how to structure a new legal framework. Should the new framework dispense with the requirement of linking emissions to ambient concentrations and instead mandate a reduction in nationwide emissions of sulfur oxides and nitrogen oxides? If so, how large a reduction in emissions should be mandated -- eight million tons? Ten million tons? Twelve million tons? How should one allocate these reductions in emissions? Should all sulfur dioxide or nitrogen oxide sources be required to reduce emissions equally? Should cars and trucks be subject to the same emissions reductions as power plant and industrial boilers? Should reduction requirements be applied to existing as well as new facilities?

What should be the timetable for these reductions? A two-step phase, as proposed in H.R. 4567 and enacted in several states? A longer or larger multi-step phase, periodically reviewed and restructured as necessary? Perhaps a single-term deadline, giving all concerned notice of the absolute requirement. Or, as EPA Administrator Thomas urged

---

[37]Colo. Rev. Stat. §§ 13-1.5-101 to 13-1.5-109 (Supp. 1986).

[38]N.J. Stat. Ann. §§ 2A:58A-1 to 2A:58A-8 (West Supp. 1986).

[39]Wisc. Stat. Ann. § 144.995 (West Supp. 1986).

in House of Representatives subcommittee testimony, is further research necessary before emissions reductions are imposed?

Who should pay for the reductions -- polluters, parts of the country benefitted by the reduction in emissions, or some combination of these two groups?

If a "polluter pays" policy is adopted, should costs be borne by the individual polluter, or equally by all the polluters within a state? If all in-state polluters pay, should cleaner utilities be required to subsidize the dirtier ones? If a polluter pays policy is discarded, what other method should be used -- an emissions tax or a generation tax? If a national tax is levied, should subsidies be provided to contain users' costs? Should the goal of H.R. 4567 to protect residential rates be maintained, or should all customers -- residential, commercial, and industrial -- be subsidized to protect them against excessive electrical rate increases? What should be the threshold level for excessive rate increases?

If these questions are answered, the steps in implementing the framework create another set of issues. What should be the extent of federal action and authority? How should the actions of EPA, the Federal Energy Regulatory Commission, the Interior Department's U.S. Geological Survey, the Agricultural Department's U.S. Forest Service, and the National Oceanic and Atmospheric Administration be coordinated? Should all administrative action be consolidated within one agency? Should the Acid Deposition Task Force serve as more than a research council? Should a standing council include nongovernmental representatives, from industrial and environmental organizations, as mandated by statute in Wisconsin?

The problem becomes further complex because fifty-one jurisdictions will be affected. What role should a state play in implementing the new legal framework? Should the framework of H.R. 4567, which gives states the full choice in determining the reduction method, be retained? Should the state's authority to choose a method be limited by mandated statutory alternatives or by administrative policy preferences? Should the current emphasis under the Clean Air Act on individual states be maintained? Or should the legislative framework move towards air quality control regions and implement reduction targets by these interstate divisions?

Until a consensus is reached on these and other key issues, Congress will continue to deliberate the passage of a bill to address the problem of acid rain; and, in the interim, legal efforts will continue to focus on Sections 115 and 126 of the Clean Air Act.

Legal Aspects of the Source–Receptor Relationship: An Agency Perspective

Charles S. Carter
Air and Radiation Division, Office of General Counsel
United States Environmental Protection Agency
401 M Street, SW
Washington, DC 20460

Let us make a very brief examination of the Clean Air Act we are dealing with now, and how effectively or ineffectively it operates to deal with the long-range transport issue. The problem is the limited knowledge that we all have on source-receptor relationship issues, and how that knowledge feeds into the regulatory process and forms our judgments on standards setting, implementation plans and so on.

I will start with some oversimplistic explanations of the statute to be sure that everyone has a basic understanding of the statute. I know that many of you have worked with it and this will be very elementary, but let's be sure that everyone understands the mechanisms that we deal with day-to-day at the agency (EPA). The key mechanism is the regulations relating to existing sources, which constitute the area that I will lump together under Sections 108, 109 and 110 of the Clean Air Act, representing a three-step process. It starts with the listing of a pollutant, based on criteria set out in 108. Once the pollutant criteria are selected, the Administrator has further the duty to establish both primary and secondary ambient air quality standards for that particular pollutant. The emphasis is, of course, on the primary standard, health. The secondary standard relates generally to public welfare effects, which include a broad range of such things as visibility and other phenomena that have been implicated in acid rain. From there, once the standard is established, the next step is the implementation of the standard itself, through the Section 110 process for establishing State Implementation Plans (SIPs).

One can see from this very elementary description that it is very complicated, and a long process to get from the selection criteria to the establishment of numerical emission limits for a particular source in a particular state to alleviate an effect back at some deposition point. I doubt that anyone who has been involved in this process would disagree that it can be a very long, tedious process. And, in particular, as we gain increased knowledge relating to the criteria pollutants, which include both sulfur dioxide and nitrogen oxides, that increased knowledge can both aid and hinder the decision-making process. Certainly, the increased knowledge can put the agency in the position to establish better standards. On the other hand, it appears to me that there can be a detrimental effect because you never arrive at a point of saying, "Oh, there's the answer; we can hone in at this point and set a standard." The diversity and range of information the agency receives can make the regulatory decision much more difficult, especially with a standard setting process like 109, which requires EPA to concentrate on a particular effect or set of effects, and then establish a standard with an adequate margin of safety.

Published 1988 by Elsevier Science Publishing Company, Inc.
Acid Rain: The Relationship between Sources and Receptors
James C. White, Editor

Particularly in an area like acid rain, where you have a range of effects about which you have scattered and incomplete information, the agency is not presented with a nice, neat basis for an informed regulatory judgment. In such a case, you must consider the ongoing SO_2 and particulate standard reviews that are underway, and are at different stages of progress. EPA proposed a fine particulate standard (PM_{10}) nearly two years ago; it has been back through the regulatory mill, and at this point a standard is much closer. On the other hand, there is no way of predicting what kind of time frame would be required to develop an acid rain standard.

The sulfur dioxide standard is presently under review and a proposal package should be out within the next year or so. Once again, this review tends to emphasize the cumbersome nature of the process. Much of this is inherent in the ambient standard process, which starts from a cause and effect relationship and works back to the source. Some have suggested that life would have been a lot simpler for those of us in the air business if Congress in its wisdom had crafted an Air Act that was more akin to the Water Act, using a technology-based approach. In that case, we might well not be here tonight having this discussion. Presumably we would have established technology-based standards for all existing sources some years ago, and I suspect that might well have obviated the need for a lot of this.

An area of litigation that some of you may or may not be aware of is related to petitions we received several years ago from the States of New York, Pennsylvania and Maine, in which they raised points regarding compliance with the ambient air quality standards and their ability to meet those standards because of emissions arising from other states. The three states alleged that emissions from certain midwestern states were preventing them from attaining and maintaining standards within their states. A regulatory decision was made two years ago, in which it was concluded that there was not sufficient evidence on the record to support the regulatory conclusions they were asking for. The SO_2 emissions from the midwestern states were not demonstrated to be causing impermissable effects in those three states, in the Administrator's judgment.

Consideration of those petitions led right back to the process I started with -- the ambient standards. The judgments that had to be made lead right back to many of the same pathways, and involve a very long, complicated cause and effect relationship, and complex modeling issues. The bottom line was that the evidence was not sufficient to support the conclusions that the three states had asked for. They, of course, challenged EPA's denial of the petitions and that case is presently pending in the U.S. Court of Appeals for the District of Columbia Circuit. At one point I had some expectation that we might have a decision on that case prior to this meeting, but obviously the court is going to let us dangle a bit longer.

There are other programs that the agency has looked at under the statute to see if there might be direct benefits or ways to achieve some direct result. But, the only other provision that is directly applicable to this sort of problem is the 115 provision, which relates to international pollution. The 115 process basically allows or authorizes the Administrator, upon receipt of certain information relating to adverse effects on another country, to draw certain conclusions, make findings, and then notify the governors of the emitting states. Very late in the Carter administration, then-Administrator Costle issued a press release and two letters to then Secretary of State Muskie and Senator Mitchell of Maine, indicating that he was in receipt of certain information from the International Joint Commission that led him to believe that adverse effects caused by acid rain were occurring in Canada, and that the sources emitting pollutants leading to those effects were, at least in part, based in the United States. He did not go any further to try to pinpoint the particular states or even the particular pollutants in the letters or the press release. He simply indicated that he was issuing instructions to the staff to undertake the next step in identifying states and pollutants, and how much emissions would have to be reduced in order to reduce the effects in Canada.

There was, undoubtedly, a basis to believe that some effects were occurring in Canada, but EPA was not in a position to make the kinds of findings that we judged that 115 required, and even less ready to initiate the sort of regulatory program to order emission reductions. After several years, a group of states and environmental groups led by the State of New York filed a lawsuit in U.S. District Court for the District of Columbia challenging our alleged failure to implement Section 115. They argued that Administrator Costle had made the requisite findings and that, as a result, the agency had a mandatory duty to proceed under Section 115. The District Judge agreed with the plaintiffs and ordered us to review a preliminary inquiry related to reciprocity. I will not discuss the inquiry, but suffice it to say, if the reciprocity obligation was fulfilled, the Judge ordered the present Administrator to move forward with notice to the appropriate governors within 180 days thereafter.

We were surprised that the Judge had ordered such summary action in the case. We immediately appealed, and I guess most of you know the result. The U.S. Court of Appeals panel, in a unanimous opinion written by then-Judge Scalia, concluded that the District Judge had gone too far, ruling that the Agency must have engaged in a rulemaking procedure for this type of finding to be binding. A simple press release and two letters issued by an Administrator simply were not sufficient to bind the agency to action in the manner the District Court found. We were somewhat surprised by the result of this case, not as to the reversal of the lower court as much as the nature of the ruling. We had not expected the panel to adopt this particular theory. Several others had been advanced that I thought more likely to justify reversing the District Court. The result was correct, but surprising.

Regarding the Court of Appeals' opinion, we are still looking into it to determine what the potential results may be in other areas outside the 115 procedure. It does have potential effects on other types of programs, and on the way EPA conducts its business in those programs.

But to re-focus on the subject here, if the agency had been ordered to go forward in some fashion with the 115 program, we would have had more difficulty with this particular program than with the ambient program. That is due, in part, to the fact that the international program is directed at precisely one thing. It is aimed at a particular country where specific findings have been made. If the agency were in a position with the kind of information and knowledge it needs to craft an intelligent approach related to acid rain effects in Canada, that is all it would be -- a program designed to alleviate the effects in Canada under Section 115. It might have incidential benefits for New England, the northeast, or other parts of the country, but it is a very narrowly focused provision that would not put us in a position to address acid rain in the broader context.

To conclude, I think that the agency is presently faced with a statute that was crafted originally in 1970, with 1977 refinements, to gauge basically localized pollution problems based on the ambient approach established in the 108, 109 and 110 process. It is just not well-suited to the kind of phenomenon we are dealing with here. There are ways to try to bend it a bit to address some of these concerns; but, in the absence of the type of knowledge that we need to act on a broad basis, an ambient approach is certainly not available now, in my judgment. And, the statute does not offer any good alternative approaches.

So we are left in a situation where we have to press on with all vigor to obtain the knowledge we need to make informed judgments on what type of regulatory program, if any, is appropriate to address the acid rain phenomenon.

Under the Clean Air Act, EPA Cannot Allocate Emission Reductions in Light of the Uncertainties Associated with Source–Receptor Relationships

Michael L. Teague
Partner, Hunton & Williams[1]
2000 Pennsylvania Avenue, NW
Washington, DC 20006

The question that I address tonight is whether the United States Environmental Protection Agency (EPA or the Agency) has authority under the Clean Air Act to allocate emission reductions to limit acid deposition in light of the uncertainties associated with source-receptor relationships. My answer examines source emissions of sulfur dioxide (SO_2) given the focus of the acid deposition debate on sulfates -- a pollutant formed in the atmosphere following the emission of SO_2.

Two regulatory vehicles in the Clean Air Act (the Act or CAA) bear on our question: attainment of the ambient standards and the control of international air pollution. Both vehicles require emission reductions at the source for the purpose of achieving benefits at an environmental receptor. Accordingly, source-receptor relationships must be understood before emission reductions can be ordered. As we will see tonight and you will explore in technical detail during your conference tomorrow, the factual record does not establish the requisite relationships. Without them, EPA cannot require or allocate emission reductions.

AMBIENT STANDARDS AND THE INTERSTATE POLLUTION PROVISIONS OF THE CLEAN AIR ACT

Following a discussion of the statutory framework for attainment of the ambient standards and for the abatement of interstate pollution, this section explains why the EPA Administrator cannot at the present time allocate emission reductions under the Act.

Legal Background

The Clean Air Act Amendments of 1970 and 1977, 42 U.S.C.A. § 7401 *et seq*. (hereinafter all citations to the Act are to the Public Law section numbers, e.g., CAA § 101), establish a combined state and federal program to control air pollution within and between states. In this program, EPA sets nationally applicable ambient air quality standards: primary standards to protect the public health, and secondary standards to protect the public welfare. CAA §§ 108 and 109. EPA has promulgated primary and secondary ambient standards for several pollutants including SO_2 and total suspended particulates (TSP). 40 C.F.R. §§ 50.4 and 50.5. The Agency has not established standards, however, specifically for sulfates -- the fine particulate pollutant that has attracted the most attention in the public debate about acid deposition.

The Act requires each state to develop an implementation plan (SIP) for "implementation, maintenance and enforcement" of the national ambient air quality standards throughout the state. CAA § 110(a)(1). The Act then directs the states to submit each SIP to the Administrator of EPA for federal review and approval. The Administrator must approve the SIP, or any SIP revision, if it meets the criteria set out in sections 110(a)(2)(A) to (K) of the Act. Train v. NRDC, 421 U.S. 60 (1975).

Section 110(a)(2)(E) of the Act establishes one of the criteria that the Administrator must consider before approving a SIP or SIP revision and "outlines the type of interstate impacts that are permissible." Connecticut v. EPA, 696 F.2d 147, 151 (2d Cir. 1982). Among other things, the EPA Administrator must assure that a SIP or SIP revision does not prevent the attainment or maintenance of an ambient standard in another state by requiring that the SIP contains:

> adequate provisions (i) prohibiting any stationary source within the State from emitting any air pollutant in amounts which will (I) prevent attainment or maintenance by any other State of any such national primary or secondary ambient air quality standard, or (II) interfere with measures required to be included in the applicable implementation plan for any other State under [the prevention of significant deterioration (PSD) or visibility protection provisions of the Act], and (ii) insuring compliance with the requirements of section 126 of [the Act] relating to interstate pollution abatement.[2]

Section 126 of the Act contains a procedural mechanism for enforcing this substantive standard regarding interstate pollution. Connecticut v. EPA, 656 F.2d 902, 907 (2d Cir. 1981). Section 126(b) authorizes any state, or political subdivision of a state, to "petition the [EPA] Administrator for a finding that any major source emits or would emit any air pollutant in violation of the prohibition of section 110(a)(2)(E)(i)." [3]

Application of the Interstate Pollution Provisions of the Clean Air Act

Over the past five years, some downwind states and environmental groups have attempted to persuade federal courts to reverse EPA approvals of SIP revisions or to mandate emission reductions on the basis of the interstate provisions of the Act. These petitioners have argued, among other things, that SO_2 emissions from sources in a neighboring state have adversely affected sulfate levels, thereby preventing the attainment of the TSP ambient standards in downwind states. These attempts have been unsuccessful. See Connecticut v. EPA, supra; Connecticut Fund for the Environment v. EPA, 696 F.2d 169 (2d Cir. 1982). These cases affirmed EPA's finding that current uncertainties associated with source-receptor relationships make it impossible for the Agency to allocate SO_2 emission reductions in order to control ambient TSP loadings. Thus, EPA cannot do so now.

In Connecticut v. EPA, petitioners sought to overturn EPA's approval of a New York SIP revision authorizing increases in the sulfur content of certain fuel oil. The United States Court of Appeals for the Second Circuit upheld EPA's final rule for three reasons. First, noting that Connecticut had met the national ambient air quality standards for SO_2, the court rejected petitioners' invitation to apply section 110(a)(2)(E)(i)(I) of the Act when there are no violations of the SO_2 ambient standards. 696 F.2d at 156.

The court next considered and rejected petitioners' claims that EPA's rule violated section 110(a)(2)(E)(i)(I) of the Act because additional particulate sulfates contributed by the approved emissions would allegedly exacerbate existing exceedances of the ambient TSP standards in Connecticut and, therefore, would "prevent the attainment" of the TSP standard in Connecticut. After determining that section 110(a)(2)(E) "seems precisely tailored to require the EPA to consider the effect of a revision of one state's implementation plan upon all national ambient air quality standards in other states",[4] the court upheld as reasonable EPA's approval of the New York SIP revision because any impact on particulate levels in Connecticut was minimal:

> where the impact upon a nearby state of another state's revision of its SIP is shown by the Agency to be so insignificant as to be fairly described as minimal, the EPA may approve that revision even where the affected state is not in compliance with the [ambient standards].[5]

Finally, the court held that EPA did not violate section 110(a)(2)(E)(i)(I) of the Act when it made no specific estimate of the impact of SO_2 emissions from the sources in New York on sulfate formation in Connecticut. The court reasoned that EPA, as yet, has no adequate model to predict such impacts; and, under the specific circumstances of this case -- the administrative record indicated that sulfate levels in Connecticut were not the product of the SO_2 emissions from the sources in New York -- the Agency reasonably concluded that it need not consider the possibility of sulfate formation. Specifically, the court stated:

> It can hardly be said that the Agency's failure to consider an effect it cannot measure constitutes a violation of the Clean Air Act.[6]

Thus, federal courts have held conclusively that, in light of the uncertainties associated with source-receptor relationships, EPA may conclude that it is unable to allocate emission reductions.

On several more recent occasions, EPA has affirmed that there is still scientific uncertainty about, and absence of a reliable technique for, modeling SO_2 transformation to sulfates. In 1984, for example, EPA considered petitions filed under section 126 of the Act by three eastern states claiming that air pollution from sources in the midwest caused, among other things, acid deposition in the petitioning states.[7] In denying the petitions, the Agency explained that air quality simulation models are not now adequate for definitive regulatory action as required under section 126.[8]

In addition, EPA's recently promulgated Guideline on Air Quality Models (Revised) (July 1986) (the Guideline)[9] confirms that adequate models do not now exist upon which the Agency may rely in allocating emission reductions to control sulfates. Appendix A of the Guideline contains summaries of refined air quality models that EPA characterizes as the "preferred" models for specific regulatory applications.[10] Notably, there is no preferred regional sulfate model in the Guideline recommended for use in resolving regulatory issues. This omission is especially significant because the Agency was aware of at least one model designed to calculate concentrations of sulfates[11] but did not recommend that model as a "refined" model for regulatory use. In promulgating the Guideline, the Agency apparently recognized that there is, as yet, no valid model for use in allocating SO_2 emission reductions to control sulfates.

And, again, in a notice of proposed rulemaking published on December 2, 1986, under section 169A of the Act (the visibility protection provisions), EPA recognized the limits of its knowledge with regard to long-range transport of pollution and declined to approve a portion of Vermont's revised SIP. 51 Fed. Reg. 43389-94 (1986). In its SIP revision submittal, Vermont asked EPA to control the out-of-state sulfate contribution to regional haze within its borders. EPA declined to approve the proposal stating that the tools necessary to address regional haze are inadequate. The Agency explained that such tools --

> are not fully available . . . . For instance, EPA does not have reference methods for monitoring nor modeling; thus, further work is needed to develop a national regulatory program.[12]

In sum, EPA has consistently found that current uncertainties associated with source-receptor relationships make it legally as well as factually impossible for the Agency to allocate SO_2 emission reductions in order to control ambient sulfates.

New Ambient Standards and Acid Deposition

EPA has authority to adopt additional ambient standards. The Agency staff has considered new ambient standards for sulfates because of acid deposition, for SO_2 to afford additional protection to exercising asthmatics, and for fine particles (principally sulfates) to control visibility. As explained below, the absence of reliable source-receptor relationships (and other factual issues) have precluded EPA from adopting a new ambient standard for acid deposition. In addition, the Agency has yet to decide that the available facts justify a new SO_2 or fine particle standard.

A Secondary Ambient Standard for Acid Deposition. As noted above, EPA has not established ambient standards specifically for sulfates.[13] The Agency staff has, however, considered the need for a sulfate standard for acid deposition.[14] While there are many factual and other concerns about an ambient standard for acid deposition, I will focus on the source-receptor issue here.

EPA may not now propose an ambient standard for sulfate deposition because, among other reasons, the Agency does not have the factual basis upon which to establish the required source-receptor relationships. Under sections 108 and 110 of the Act, proposal of ambient standards must await development of information on how to implement the standards. This information includes techniques for identifying which sources are causing exceedances of the proposed standard and methods for controlling those sources that will alleviate the exceedances. *See* CAA § 108(b). This information, as the legislative history makes clear, is "need[ed by the states] . . . to develop meaningful programs for implementation of ambient air quality standards . . . ."[15]

The source-receptor relationships for sulfates now available to the EPA Administrator do not permit him to issue to the state the detailed information required by the Act.[16] Because unambiguous attribution of ambient sulfates to a specific source is impossible, EPA does not have the factual record necessary to propose a secondary ambient standard for sulfates to control acid deposition.[17]

SO2 Controls and Other New Ambient Standards. In its evaluation of the existing ambient standards for SO_2, EPA is considering whether a new short-term standard to protect asthmatics while exercising is needed. Alternative short-term standards could produce results that range from substantial additional SO_2 emission controls to no deviation from the status quo.

EPA has also been examining the need for a secondary ambient standard for fine particles to protect visibility. To date, the Agency's staff has focused its attention on the sulfate contribution to regional haze and possible SO_2 controls if such a secondary standard were to be adopted.

Some argue that concerns about acid deposition justify adoption of a stringent one hour primary ambient standard for SO2 or a fine particle secondary ambient standard so as to force additional SO_2 emission reductions. The Clean Air Act, however, makes clear that ambient standards must be based on benefits to specific environmental receptors, for example, to public health for primary standards and visibility for secondary standards.[18] Vague concerns about acid deposition are inadequate support for ambient standards for health and visibility. In other words, EPA has no authority to adopt acid deposition controls through the back door.

INTERNATIONAL AIR POLLUTION

Emissions from sources within the United States that cause or contribute to air pollution that endangers the public health and welfare of the citizens of another country can be regulated under section 115, the Clean Air Act's international air pollution provision. Section 115 is similar to the ambient standards provisions of the Act because under section 115, emission reductions for a particular source must be based on the need to achieve environmental benefits at a specific receptor.

Under section 115, the EPA Administrator is authorized to notify a state governor that air pollutants emitted from sources in his state may endanger public health or welfare in a foreign country.[19] Section 115, therefore, imposes upon EPA a substantial evidentiary burden. Before a finding that air pollutants emitted "in the United States" endanger public health or welfare in a foreign country can give rise to a formal notice to a state governor, the Administrator must identify the origin of the substances.[20] This task is enormously difficult for all pollutants and especially for sulfates, not only because numerous sources across the country may contribute to acid deposition, but also because their long-range transport may make determination of their origin impossible. Thus, EPA needs valid and reliable source-receptor relationships to assemble a complete, accurate factual record under section 115.

As explained in papers filed by EPA in the section 115 litigation, no such techniques exist now. Charles L. Elkins, then acting Assistant Administrator for air programs at EPA, filed an affidavit explaining that models are not now reliable enough to implement section 115, that the Agency does not have adequate experience in designing the control programs that are needed, and that the Agency could not now allocate emission reductions in the defensible way required by section 115.[21]

Thus, in the international air pollution context, the Clean Air Act calls for emission reductions at the source to benefit environmental receptors. Hence, source-receptor relationships must be known before EPA can allocate emission reductions. Because the requisite relationships have not been established between SO_2 and sulfate deposition, the Agency cannot now allocate emission reductions to control the impact of sulfates in foreign countries.

LACK OF NEED FOR EMISSION REDUCTION ALLOCATIONS

In testimony before Congress as well as the federal courts, EPA has indicated that no need can now be demonstrated for additional SO_2 controls based on sulfate effects. In his Affidavit to the United States District Court for the District of Columbia, for example, Charles Elkins stated:

> At present, it is impossible, based on the results of [the on-going ten-year research program on acid deposition] and despite the four years and $132 million dollars already invested, to draw sound and scientifically defensible limits between acid deposition and environmental damage.[22]

As a result, Mr. Elkins told the court that EPA could not justify the imposition of emission reductions to control acid deposition.

Similar conclusions have been expressed by EPA Administrator Lee M. Thomas in recent testimony before Congress. On June 20, 1986, for example, Mr. Thomas told the United States House of Representatives that --

> [EPA does] not believe that the current state of knowledge can sustain any judgment with respect to the level of emission reductions needed to prevent or eliminate [the threat of acid rain] damage . . . . [EPA does] believe, however, that current research is providing answers to these questions and that little or no additional damage will occur while we are waiting for this necessary information.[23]

As Mr. Thomas told the United States Senate earlier, the National Acid Precipitation Assessment Program (NAPAP)[24] is hard at work developing the necessary information to characterize the relation between acid deposition and environmental damage.[25] The early results from this comprehensive federal research program suggest that initial claims made about acid rain are either unfounded, incorrect or exaggerated.[26] In fact, research to date shows that acid rain controls cannot reasonably be expected to produce environmental benefits commensurate with the cost of those controls. As Mr. Thomas told Congress:

> Given the present uncertainty that . . . any acid rain control will prevent or eliminate observed damage to critical resources, and the certainty that it will impose enormous costs, we believe <u>enactment of acid rain legislation at this time would be inappropriate</u>.[27]

CLOSING

Thank you for the opportunity to address you tonight. I hope that my comments on the legal and factual requirements for emission reduction allocations to control acid deposition under the Clean Air Act have added to your understanding of why EPA has declined to make such allocations to date.

FOOTNOTES

[1] I gratefully acknowledge Maida Lerner's significant contribution to the preparation of this paper.

[2] CAA § 110(a)(2)(E)(emphasis added).

[3] After receiving a petition under section 126 and holding a public hearing, the EPA Administrator must either find a violation of section 110(a)(2)(E)(i) or deny the petition. If the Administrator finds a violation, the statute provides that the offending source must shut down or meet strict deadlines for compliance with new emission limitations. CAA § 126(c).

[4] 696 F.2d at 163.

[5] 696 F.2d at 165 (emphasis added). In reaching this interpretation, the court noted "the language of [§ 110(a)(2)(E)(i)(I)] seems to contemplate a standard which would prohibit SIP revisions only if the emissions they permit

would in and of themselves prevent a nearby state from attaining the [ambient standards]." 696 F.2d at 164.

[6] 696 F.2d at 165. See also, Connecticut Fund for the Environment v. EPA, 696 F.2d 169, 177 (2d Cir. 1982) (where the United States Court of Appeals for the Second Circuit held that the issue of statutory construction of section 110(a)(2)(E) was effectively moot with regard to the interstate impacts of sulfate in the absence of any reliable technique for modeling the SO_2 transformation to sulfates).

[7] 49 Fed. Reg. 48152 (1984) (Final Determination under section 126 of the Clean Air Act, Interstate Pollution Abatement); 49 Fed. Reg. 34851 (1984) (Proposed Determination under section 126 of the Clean Air Act, Interstate Pollution Abatement).

[8] EPA also indicated in that proceeding that Congress did not intend that section 126 petitions be used to control a regional pollutant like sulfates. 49 Fed. Reg. at 48154. EPA stated:

> In enacting Section 126, Congress was not attempting what it refrained from doing in the Act as a whole, i.e., to establish a comprehensive allocation system for air resources. Rather it was trying to cope with those instances where one State makes it impossible for another to meet its mandatory obligations under the Act. To interpret section 126 more broadly would authorize, indeed require, the Administrator to redistribute air resources among the states which, surely, lies beyond the mandate of the Agency.

49 Fed. Reg. at 34858 (col. 2).

[9] According to EPA, the Guideline:

> recommends air quality modeling techniques that should be applied to [SIP] . . . revisions for existing sources and to new source reviews, including [PSD] . . . . It is intended for use by EPA Regional offices in judging the adequacy of modeling analyses performed by EPA, State and local agencies and by industry . . . . It serves to identify . . . those techniques and data bases EPA considers acceptable.

Guideline at 1-1.

[10] EPA notes that each of these preferred models has been "subjected to a performance evaluation using comparisons with observed air quality data" and "may be used without a formal demonstration of applicability." Guideline at A-1.

[11] See Guideline at B-43 (where EPA describes the "MESOPUFF II" model).

[12] 51 Fed. Reg. at 43392. EPA took a similar position in a recent judicial proceeding brought pursuant to section 115 -- the international pollution provisions contained in the Act -- the subject of the next section of this paper. Thomas v. State of N.Y., 802 F.2d 1443 (D.C. Cir. 1986) (United States Court of Appeals for the D.C. Circuit overturning a lower court ruling that EPA had to initiate and complete, by a date certain, a proceeding to control transboundary pollution under section 115 of the Act). In an Affidavit to the United States District Court dated September 9, 1985, Charles Elkins, then the Acting Assistant Administrator of EPA's Office of Air and Radiation, affirmed that

> mathematical models that simulate the movement and chemical transformation of pollutants in the atmosphere over long distances . . . do not yet produce results that are reliable enough to be used in developing regulatory programs.
>
> * * * *
>
> [Thus] emission reduction obligations cannot be allocated among states until the tools needed to calculate them . . . have been developed and the problem has been fully analyzed through their use.

[13] The current ambient standards for SO_2 provide, however, for substantial control of a sulfate precursor.

[14] U.S.E.P.A, Environmental Criteria and Assessment Office, Pub. NO. 600/8-82-029b, Air Quality Criteria for Particulate Matter and Sulfur Oxides, Vol. 2, Ch. 7 ("Acidic Deposition") (1982).

[15] S. Rep. No. 91-1196, 91st Cong., 2d Sess. 9 (1970); see CAA § 110 (a)(2)(B) and (E). And under the statutory scheme, EPA may not promulgate a standard with the hope of developing implementation procedures in the future because steps to attain and maintain ambient standards must be executed consistent with

tight statutory deadlines. Under section 108(b), for example, EPA must publish source identification techniques and control methods at the time that the Agency proposes an ambient standard. Under § 110, the states have only nine months to submit a SIP following promulgation of an ambient standard; and EPA has but four months to act on the SIPs. CAA § 110(a)(1) and (2). Reliable source-receptor relationships must be available to support this SIP development and review process.

[16] The discussion above summarizes EPA's findings on this issue.

[17] This analysis ignores the issue of the need for such a standard. The "need" issue is discussed below.

[18] CAA §§ 108(a)(1)(A), 109(b)(1) and (2). See also H. Rep. No. 91-1783, 91st Cong., 2d Sess. 4 (1970).

[19] CAA § 115(a). As noted above, this proceeding is similiar to that authorized under section 110 of the Act. Under section 110, a SIP is required to include "emission limitations, schedules, and timetables for compliance with such limitations" for sources within that State in order to insure the "attainment and maintenance" of the national ambient air quality standards. CAA § 110(a)(2)(B). If the standards are exceeded, the Act contemplates that the SIP will be revised. CAA § 110(a)(2)(H). Similarly, section 115 is to be implemented through the SIP revision process. CAA § 115(b).

[20] The Act provides that the Administrator of EPA shall give formal notification of such effects "to the Governor of the State in which such emissions originate." CAA § 115(a)(emphasis added).

[21] State of New York v. Thomas, supra, (Elkin's Affidavit). See note 12 above. This position is clearly consistent with that taken by the Agency in the earlier judicial proceedings discussed in the first section of this paper as well as in the more recent rulemakings under sections 126 and 169A.

[22] State of New York v. Thomas, supra, (Elkin's Affidavit).

[23] Hearings on H.R. 4567 before the Subcommittee on Energy Conservation and Power of the United States House of Representatives Committee on Energy and Commerce (June 20, 1986) (Statement of Lee M. Thomas, EPA Administrator, 2).

[24] NAPAP is the coordinating body for acid rain research ongoing at EPA and other federal agencies including the

Departments of Energy, Health and Human Services, Agriculture and Interior.

[25] Hearings on Acid Rain before the United States Senate Committee on Environment and Public Works (December 11, 1985) (Statement of Lee M. Thomas, EPA Administrator).

[26] CHK for cite to NAPAP annual report.

[27] Hearings on H.R. 4567 before the Subcommittee on Energy, Conservation and Power of the United States House of Representatives Committee on Energy and Commerce (Statement of Lee M. Thomas, EPA Administrator (June 20, 1986) (emphasis added).

Scientific Uncertainty, Agency Inaction, and the Courts

David R. Wooley
Assistant Attorney General, New York State Department of Law
Albany, NY 12224

ABSTRACT

States and citizen groups are using court actions to force the federal government to act on the problem of acid rain.[1] The uncertainty in the science of source-receptor relationships is being used by the agency to resist demands for action and to obscure EPA's obligations under existing law. Courts, however, generally will not excuse failure to carry out statutory commands on this basis. The Clean Air Act contemplates that decision making necessary to protect health and the environment will often have to be carried out in the face of uncertainty about causes and effects. Acid rain legislation pending in Congress takes this same approach. Certainty in the science of source receptor relationships is largely unattainable. What is attainable, and fairly likely in the near future, is the imposition of acid rain abatement requirements by Congress or the courts. Government and private research efforts should be oriented to that reality. Attempts to delay agency action should be abandoned in favor of an immediate effort to identify a solution based on the best available data and analytical tools.

INTRODUCTION

This symposium provides an important opportunity to define what we know and do not know about long-range transport of air pollution. It will help to identify the areas most in need of further research, to explore significance of the remaining uncertainties, to test the strength of the current theories, and to work toward greater consensus on what to do about the problem of acid rain. Which abatement procedure will be most effective? Which pollutants should be reduced and in what proportion? Where should the reductions occur? These are some of the vital questions on which we do not yet have completely definitive answers.

But, while you are engaged in this process, I urge you to keep in mind that the existence of scientific uncertainty on these questions does not necessarily mean that abatement action will be postponed. The quest for certainty in this field will not end in the near future, and there may always be unanswered questions. The answers we do achieve may only create new questions of their own. A requirement for abatement action is likely to be imposed long before there is an opportunity to resolve much of the remaining uncertainty.

Acid Rain: The Relationship between Sources and Receptors
James C. White, Editor

Abatement action will occur either as a result of the enactment of an acid rain control bill by Congress, or by reason of court orders directing EPA to act on the basis of its existing statutory obligations. In either event, as a matter of law, scientific uncertainty or the desire for more complete answers will not be sufficient to excuse EPA inaction. The object of research in the short term, therefore, should be the identification of a practical emission abatement plan.

I realize that for some scientists this must sound like heresy. But on the issue of acid rain the scientist and the regulators must work together. To do so effectively they must recognize that their tasks are fundamentally different, particularly in the way they address scientific uncertainty. To paraphrase Judge Bazelon, scientists seek to conquer uncertainty, while regulators must act in spite of it. If anything, the scientist is more likely to overemphasize uncertainty, while the regulator cannot afford the luxury of withholding judgment if he is to carry out Congressional commands for action against hazards to health and the environment.[2] This perceptual difference is particularly important for atmospheric scientists and acid rain regulators who are acutely aware that virtually every judgment involves intricately interacting uncertainties in chemistry, physics, meteorology, statistics and computer sciences. Scientists must, as Judge Bazelon says, "press onward upon the line between the known and the unknown," but in doing so they must not seek to impose the scientific view of uncertainty upon the regulatory process. Similarly, the regulator must disclose and be aware of uncertainty, but not use it as an excuse for inaction or to second guess the mandates of Congress.

Thus, it is important to recognize that the great scientific debate over the causes and effects of acid rain occurs within a legal framework, and that the principles which guide the regulators may be as important as the science itself in determining when society ultimately responds to the problem.

This article explores some of the legal principles which affect decision making on this controversy, particularly as they relate to uncertainties in the science. Some of the principles come directly from Congress. Others have evolved from traditions of the courts. Although the executive branch is in a particularly powerful position to apply, defy, or make over the traditional ways of applying law to science in decision making, it may be overruled by the courts or the Congress. This article will discuss some of these principles in the context of a case now awaiting decision in the United States Court of Appeals in Washington. The outcome of that litigation could mean that long-range source receptor modeling will finally be used in regulation and control of acid rain, in spite of the remaining uncertainties, and EPA's opposition.

The analysis begins at the place where the journeys of all lawyers, judges and policy makers must start: the words of the law -- in this case the federal Clean Air Act ("Act").[3]

THE ACT

Since 1980 a coalition of northeastern states, environmental groups, Canadian governments, and private individuals have attempted to use provisions of the federal Clean Air Act to force reluctant members of the executive branch to order reduction in emissions which cause acid rain.

The Act provides numerous opportunities for acid rain litigation efforts. It has a broad "citizen suit" provision which enables individuals and states to compel the federal government to carry out "non-discretionary" duties imposed by Congress on the EPA. It has liberal "judicial review" provisions which insure that all final agency actions, whether discretionary or non-discretionary, are carried out in a manner that is based upon reasonable apprehension of the factual issues and compliance with legal requirements.[4]

The Act is also a comprehensive law that requires remedies for virtually all types of outdoor air quality problems.[5] Some provisions prescribe specific measures for particular pollution problems.[6] However, most of the provisions of the Act do not specifically address any individual pollution or source types. Rather, the Act establishes generic and flexible mechanisms to remedy any adverse effect on public health and welfare from air pollution.

Among these generic provisions several are relevant to acid rain. These include provisions for establishing national ambient air quality standards, and the interstate and international air pollution abatement sections.[7] Each of these provisions follows a pattern which has become standard in all environmental legislation in the United States Code. Each provides that whenever the EPA Administrator makes a factual determination that emissions of pollution are causing harm to public health and welfare, he <u>shall</u> take actions leading to abatement.

Under §§ 108 and 109 the recognition of a pollutant's harm to health and welfare requires EPA to establish national ambient air quality standards, which in turn require the writing and implementation of emission control plans to achieve those standards.[8] In the interstate section the finding that one state emits pollutants which prevent or interfere with a downwind state's ability to attain these standards raises an obligation on EPA's part to compel emission reductions in the upwind states.[9] Similarly, under the international air pollution abatement section, when the Administrator finds that U.S. emissions are causing harm to another nation, the Administrator <u>shall</u> compel responsible states to reduce their emissions.[10]

A critical principle underlying all of these provisions is that scientific uncertainty will generally not excuse the agency from either making the threshold factual findings, or in carrying out the required abatement actions. Congress was fully cognizant that there would be limitations and uncertainties in the data available to the Administrator in making decisions under the Act.[11] The legislative history to the 1977 amendments to the Act unequivocally demonstrates that Congress concluded that prompt though imprecise regulation is preferable to no regulation at all. The legislative history quotes and approves of language from a 1976 decision of the United States Court of Appeals for the District of Columbia Circuit. There Judge Leventhal stated:

> Agencies unequipped with crystal balls and unable to read the future are nonetheless charged with evaluating effects of unprecedented environmental modification, often on a massive scale. Necessarily they must deal with predictions and uncertainty [and] with developing evidence . . . Sometimes, of course, relatively certain proof of danger or harm can be readily found. But, more commonly, theory long precede[s] certainty. Yet the statutes -- and common sense -- demand regulatory action to prevent harm, even if the regulator is less than certain that harm is otherwise inevitable.[12]

The practicality and wisdom of this approach has been borne out in the experience with EPA use of short-range modeling to attain the national ambient air quality standards in the immediate vicinity of pollution sources. Early on it was recognized that to attain the national ambient air quality standards it was necessary to utilize some model to relate emission reductions to ambient air quality improvement. When the first state plans were written and approved the agencies were forced to rely on relatively crude assumptions, such as the so-called "roll back models," involving a straight line or one-to-one ratio between emission rates and ambient concentrations. Nevertheless the courts consistently upheld EPA's methods on the grounds that the use of uncertain data is necessary if the agency is to perform its statutory function. The courts found that the Act was satisfied so long as the data and methods chosen were the best that is feasibly available.[13]

Today, of course, short-range modeling has been greatly improved, but we still talk about an expected range of error of a factor of two.[14] Thus, in looking at the state of long-range air pollution modeling today we must remember that accuracy has always been a relative term, and that the Act simply does not allow EPA to delay regulatory action while it purports to pursue the holy grail of scientific uncertainty in a field which is inherently uncertain.

Unfortunately, EPA does not agree. Although it has generally acknowledged the existence of significant harm caused by long-range transport of acid rain pollutants, EPA has refused to act upon these findings under its existing statutory mandate. EPA excuses its inaction with the claim that long-range modeling techniques are insufficiently accurate for regulatory use in tracing the origin and impact of air pollution over long distances.[15] This attitude has lead to several legal challenges under the "citizen suit" and judicial review provisions of the Act. These legal challenges involve judicial principles which also affect the practical impact of the Clean Air Act mandates.

IN THE COURTS

The problem of agency inaction in the face of statutory commands is not unique to the acid rain issue. We live in an era in which the executive branch favors less governmental intrusion into the private sector, fewer regulations, and a reduced role for federal agencies. This attitude clashes loudly with the spirit and purposes of many of the remedial environmental statutes enacted in prior years and seems particularly inappropriate in the case of an interstate and international air pollution problem that cannot be addressed effectively by local or state government. In the past six years the federal courts have been confronted with an increasing number of cases seeking to force administrative agencies to take regulatory action on issues involving complex scientific problems.[16]

In response, administrative agencies have developed a number of arguments to defend inaction. One of the most common defenses combines notions of scientific uncertainty with traditional principles of administrative law which recognize limited ability of courts to second guess technically complex factual determinations of administrative agencies.

It is well established that courts owe a certain amount of deference to factual determinations and legal interpretations of administrative agencies. This is particularly true when made within the agency's realm of technical or scientific expertise.[17] Thus in reviewing an agency's regulation the courts will not overturn agency determinations and action where there is supporting evidence and the agency's action is reasonable.[18] In other words, where an agency takes affirmative action, that action will not generally be overturned simply because there are uncertainties in the evidence or because other plausible explanations or conclusions are possible. In effect, the role of the scientific expert is preserved from intrusion by less knowledgeable lawyers and judges. And yet the court review must be deep enough to uncover obvious abuses in which the government has no reasonable factual basis for its judgment and has ignored scientific opinion and consensus from inside or outside the agency.[19]

In recent years, however, the government has attempted to expand this limited "deference concept" to justify agency <u>inaction</u> as well. Initially, such efforts generally failed. The agency's view of the adequacy of scientific knowledge is less likely to be accepted where the inaction it supports collides with a Congressional command that steps be taken to protect the public from harm. Deference may be appropriate where an agency has made a choice based on its technical expertise, but not where it has essentially refused to exercise its judgment. It has not been difficult to convince courts of the common sense principle that decisions and actions needed to protect the public nearly always, in the environmental field, have to be taken on the basis of inconclusive or scanty data. It has been enough to show that Congress was especially clear on this point when it enacted the 1977 amendments to the Clean Air Act. Thus the case law in the 1970's and early 1980's generally rejected the use of the deference principle to allow agencies to avoid statutory commands on the grounds of scientific uncertainty.[20]

Nonetheless, despite the clear precedent to the contrary, EPA has continued to press the scientific uncertainty rationale upon the courts as a defense to citizen suit and judicial review petitions that seek to force EPA action. The history of our litigation under the interstate air pollution section of the Act illustrates the continuing controversy.

INTERSTATE AIR POLLUTION CASE

The Clean Air Act establishes an obligation on EPA to determine that each state air pollution control plan is adequate to prevent pollutants from flowing across state lines in such amounts as to cause violations of national ambient air quality standards in other states. 42 U.S.C. § 7410(a)(2)(E). The Act also establishes a petition process whereby states can petition EPA for a determination whether or not upwind states have adequately controlled their emissions as necessary to prevent such prohibited interstate air pollution. 42 U.S.C. § 7426.

In 1981 and 1982 New York, Pennsylvania and Maine filed § 126 petitions claiming that inability to attain the sulfur dioxide and particulate standards and to improve visibility in the petitioning states was caused by sulfur dioxide and sulfate particles arising from poorly controlled emissions in upwind states. The states asked EPA to find that named midwestern states' emissions were responsible for these problems. Section 126 imposes a 60-day deadline for EPA to answer the petition, and states that if a finding is made in favor of a petitioning state then the responsible upwind states must reduce their emissions according to directives to be established by EPA.

Initially EPA took the position that it could indefinitely delay any ruling on the petitions. In 1984, the states filed a citizen suit under § 304 of the Act to force EPA to either make the requested finding, or to deny the petitions upon a finding that upwind states' air pollution plans are adequate to prevent impermissible interstate air pollution.[21] Although EPA acknowledged that it had been considering the evidence for over two years, it claimed complexity of the "chemistry and meteorology of long range transport of air pollutants, emissions data and trends" justified further delay, and that the court should "accord deference to the agency" by refraining from imposing any deadline for decision making on the agency.[22]

The District Court rejected this argument. On October 5, 1986 the court held that it was "unseemly" for the Administrator to assert that he is vested with discretion to balance the need for prompt regulation against the need for greater technical information. It found that Congress clearly intended to establish prompt regulations and control over the "serious problem" of interstate air pollution and that Congress "did not expect [EPA] to resolve every potential problem" before making decisions on § 126 petitions.

> By creating the § 126 petition process, Congress sought to establish a means of protecting citizens and the environment from harmful effects of air pollution originating outside their home state. Defendant's delay in following his statutory mandate has compromised this process.[23]

The court ordered EPA to issue a decision on the § 126 petitions within 60 days. EPA complied by denying the petition.[24] In doing so, however, it refused to determine whether or not the upwind states air pollution plans were adequate to prevent interstate pollution or whether impermissible interstate pollution was in fact occurring. Rather, EPA denied the petitions on the grounds that it simply did not know enough in order to make these determinations.

On appeal the northeastern states argued that, regardless of uncertainties in the data, EPA is required to make the determinations on the basis of the best information available to carry out the intent of the Act. They pointed out that EPA made no effort to rebut the evidence supporting the petitions and had admitted that the states' modeling represented the state-of-the-art knowledge on long-range transport of sulfur dioxide and sulfate pollution. The states concluded that under these circumstances EPA must make findings and take action on the basis of that information.[25] EPA again defended its refusal to make the determinations on the now familiar claim that the science of long-range modeling does not yet give "definitive" or completely reliable results. This blanket refusal to use long-range modeling is absurd. New York exhaustively documented EPA's ability to model long-range impacts of sulfur dioxide and sulfate pollution. EPA's own guideline specifically lists long-range pollution models available for use in special circumstances. EPA's own brief showed that its contractors had developed "a model with all components complete, running on real data sets" which, "will incorporate all known important processes dealing with emissions, transport and transformation of chemical processes . . .". New York argued that these facts, along with the modeling it presented through its expert Dr. Perry Samson, required EPA to either credit the evidence presented or come forward with modeling results of its own to confirm or contradict the petitioner's case. The case was argued in the United States Court of Appeals for the District of Columbia Circuit more than a year ago, and a decision is expected soon.

If EPA's effort to grossly enlarge the principle of judicial deference succeeds, then there will be even greater pressure for Congress to enact an acid rain bill which will leave EPA no excuse for further inaction. Indeed it was frustration with inaction on the part of EPA and its predecessor agency which promoted passage of the 1970 and 1977 Clean Air Act amendments.[26]

The states are hopeful that the court will take this opportunity to again reject the notion that an agency's desire for certainty or conclusive evidence allows it to ignore a legislative command for action to protect health and welfare.[27] If EPA's argument fails we may soon see a court decision which requires EPA to take the best available source/receptor evidence and translate it into an abatement program under existing law. In either event the scientific community must be ready to provide the government and the other litigants with its best estimation as to the causes and effects of acid rain and as to the effectiveness of various corrective measures.[28]

CONCLUSION

Scientists, law-makers and judges must understand that scientific uncertainty and regulatory action are not incompatible concepts. Rather it is the interaction between the two that is the cutting edge of decision making on environmental policy. While it is important to know what the knowledge gaps are, it is a mistake to emphasize them to the point of regulatory paralysis. EPA and the courts should accept the principle that if there is a statutory responsibility to take action, then the agency must do so based on the best available information and analytic tools. EPA should abandon efforts to convince courts to keep hands off scientific issues simply because they are technically complex, or not definitively understood. To justify inaction on these grounds affords inadequate protection for the health and safety of the public and dangerously expands the power of the executive branch over the prerogatives of the Legislature.

In the coming year we are likely to see significant legal developments affecting acid rain. If Congress acts, it will undoubtedly continue the tradition of requiring abatement action by EPA regardless of uncertainties in the science. Court decisions could also force EPA to proceed to develop and implement an abatement plan. Diplomatic pressures on the administration to take action will increase. These events will intensify the need for improved data bases and analytic tools by which decision makers can choose the best solution to this problem. The emphasis should be on finding the best, rather than the perfect; to achieve general accord rather than unanimity. The scientific message to the public and the regulators should be action oriented rather than an abstracted preoccupation with uncertainty. This is the best way to ensure that the scientific community's voice will be heard.

FOOTNOTES

1. The author served as lead counsel for plaintiffs or petitioners in several of these court actions. The author is a graduate of Rutgers College and Rutgers University School of Law, and is a member of the bar in four states and numerous federal courts.
2. David L. Bazelon, Science and Uncertainty: A Jurist's View, 5 Harvard Environmental Law Review 209, 210-211 (1981).

3. 42 U.S.C. § 7401 et seq.
4. 42 U.S.C. §§ 7604, 7607.
5. Chevron U.S.A., Inc. v. National Defense Counsel, 104 S. Ct. 2778, 2785 (1984).
6. See, 42 U.S.C. §§ 7421, 7450, 7571.
7. 42 U.S.C. §§ 7408, 7409, 7410(a)(2)(E), 7426, 7415.
8. 42 U.S.C. §§ 7408, 7409.
9. 42 U.S.C. § 7426.
10. 42 U.S.C. § 7415.
11. House Report No. 294, 95th Cong. 1st Sess., 46-48, reprinted in 1979 U.S. Code, Cong. and Admin. News at 1124.
12. Ethyl Corp. v. EPA, 541 F.2d 1, 6, 25 (D.C. Cir) (en banc) cert. denied, 426 U.S. 941 (1976). See also, Lead Industries Ass'n. v. EPA, 647 F.2d 1130, 1154-55 (D.C. Cir. 1980); Environmental Defense Fund v. EPA, 598 F2d 62, 81 (D.C. Cir. 1978).
13. Texas v. EPA, 499 F.2d 289, 301, 318 (5th Cir. 1974), cert. denied 427 U.S. 905 (1976); Cleveland Electric Illuminating Company v. EPA, 572 F.2d 1150, 1156 (6th Cir. 1978); Alabama Power v. Costle, 636 F.2d 323, 387-88 (D.C. Cir. 1979).
14. Guideline on Air Quality Models (Revised), United States Environmental Protection Agency, July, 1986, EPA-450/2-78-027R, p. 10-3.
15. See, Brief for Appellant in Thomas v. New York, D.C. Circuit Nos. 85-5970 and consolidated case, pp. 38.
16. Boyer and Meidinger, Privatizing Regulatory Enforcement: A Preliminary Assessment of Citizen Suits Under Federal Environmental Laws, 34 Buffalo Law Review No. 3, Fall 1985, pp. 833, 863-864.
17. Baltimore Gas and Electric Co. v. NRDC, 462 U.S. 87, 103 (1983). See generally, Stever, Deference to Administrative Agencies in Federal Environmental Health and Safety Litigation - Thoughts on Varying Judicial Application of the Rule, Western New England Law Review, 6:35 (1983).
18. Small Refiner Lead Phase-down Task Force v. EPA, 705 F.2d 506, 520-1 (D.C. Cir. 1983).
19. Citizens to Preserve Overton Park v. Volpe, 401 U.S. 402, 415 (1971) ("thorough, probing, indepth review"); Alabama Power v. Costle 636 F.2d 323, 359 (D.C. Cir. 1979).
20. California v. Watt, 668 F.2d 1290, 1312 (D.C. Cir. 1981); NRDC v. Herrington, 768 F.2d 1355, 1400-1403, 1415 (D.C. Cir. 1985); New York v. Gorsuch, 554 F. Supp. 1060 (S.D. N.Y. 1983); Sierra Club v. Gorsuch, 551 F. Supp. 785 (N.D. Cal. 1982); Illinois v. Gorsuch, 530 F. Supp. 340, 341 (D.D.C. 1981).
21. State of New York, et al. v. William Ruckelshaus, Civil Action No. 84-0853, United States District Court for the District of Columbia. New York, Pennsylvania and Maine were joined by New Hampshire, Vermont, Massachusetts, Rhode Island, Connecticut and New Jersey in the District Court case, and in the later appeal from the denial of the § 126 petition. See footnote 25, infra.
22. Defendants' Memorandum In Opposition to Plaintiffs' Motion for Summary Judgment and In Support of Defendants' Motion to Dismiss and For Summary Judgment, in State of

New York v. William Ruckelshaus, D.D.C. Docket No. 84-0853, pp. 6, 10.

23. Memorandum Opinion, State of New York, et al. v. Ruckelshaus, 21 E.R.C. [BNA] 1721 (Oct. 5, 1984), pp. 6-10.
24. 49 Federal Register 48152-48157 (December 10, 1984).
25. Brief for Petitioners, in State of New York v. U.S.E.P.A., (D.C. Cir. Nos. 84-1592, 85-1082), March 18, 1985, pp. 14-15, 18-39.
26. Train v. NRDC. 421 U.S. 60, 63-65 (1975); Connecticut v. EPA, 656 F.2d 902, 906 (2d Cir. 1981); House Report, supra at 135-136, 151, 330, U.S. Code Cong. News (1977) at 1214, 1230, 1409; S. Rep. No. 127, 95th Cong. 1st Sess. 41-42 (1977).
27. There is another pending case in which EPA has attempted to defend its inaction on acid rain with the scientific uncertainty claims. Environmental Defense Fund, et al. v. Thomas, United States District Court for the Southern District of New York, Civil Action No. 85-9507. The case is based on Sections 108 and 109 of the Clean Air Act and constitutes an attempt by environmental groups and northeastern states to force EPA to revise the National Ambient Air Quality standards for sulfur dioxide and particulates in a way which would force reductions in acid rain. The plaintiffs argue that the Act imposes a non-discretionary duty upon EPA to establish national standards which address each of the "health" and "welfare" effects of air pollutants which are regulated. It was shown that the existing National Ambient Air Quality standards for these pollutants were set without consideration of acid rain effects. Through EPA's own statements, the plaintiffs showed that emissions of these pollutants are the primary cause of a variety of effects which fall under the definition of "welfare." EPA has defended the lawsuit by arguing that it cannot be compelled to set acid rain based ambient air quality standards because the information about its causes and effects are not sufficiently complete. EPA is currently in the process of revising the National Ambient Air Quality standards for these pollutants, but has specifically refused to consider acid rain effects. The District Court has been asked to order EPA to do so. The result may be an order to establish national standards which, when implemented, would remedy acid rain by compelling substantial reductions in sulfur dioxide emission reduction. All papers necessary to a court decision in that case were submitted during the Spring of 1986, and a decision is expected soon.
28. Such a result would be consistent with other recent cases rejecting the scientific uncertainty defense. Natural Resource Defense Counsel v. Herrington, 768 F.2d 1355 (D.C. Cir. 1985) (appliance efficiency standards); State of New York v. Thomas, 613 F. Supp. 1472 (D.D.C. 1985) rev'd 802 F.2d 1443 (D.C. Cir. 1986) (International Air Pollution Abatement provisions of Section 115.) In the latter case the District Court first held that EPA had made findings that United States emissions were causing harm to Canada. It then addressed EPA's argument that Section 115 imposed no mandatory obligation to take abatement action because the decision about how and when

to regulate acid rain "requires the fusion of technical knowledge and skills with judgment which is the hallmark of duties which are discretionary." The District Court rejected this argument. Its opinion was later reversed by the Court of Appeals, but on other grounds. Thomas v. New York, Opinion of September 18, 1985, D.C. Cir. No. 85-5970 and consolidated cases.

Questions and Answers following Legal Aspects Symposium

Wednesday, December 3, 1986

Q:

In my opinion, even if we had perfect knowledge on the source-receptor relationship, there are still some uncertainties in terms of whether we should implement a program that is going to cost so much money to be paid by the taxpayers or the ratepayers. In view of the certainty and the uncertainty of the effects of pollutants in the environment, I ask whether or not the reductions are required to protect our environment over the long term and whether or not there is time to develop new technology to be put in place as new powerplants come on line. Is the problem so severe that in twenty years we might not get to the same place without acid rain legislation, but at a great savings to many of our citizens? That's an issue I'd like your comment on.

A: (Wooley)

I think that virtually everything the government does runs into that question. There's a great risk that it wouldn't do any good for the government to spend $25 billion over five years to develop the Strategic Defense Initiative. That's risky, that's certainly very uncertain. With acid rain, some people are going to be subjected to costs for the benefit of protecting the people's environment downwind. The Northeast has taken the position that we should help pay for this. We supported the initial cost sharing provisions in the bills back in 1984, when it wasn't such a popular idea in the northeast. We are concerned; we don't want people to suffer outrageous rate impacts; we don't want to see lots of coal miners thrown out of work. There are ways to mitigate those damages and share the costs around the country. This should be done, and the best way to do this is through the enactment of a law because, under existing law, there is no provision for cost sharing. If we succeed, people will start to pay attention and say, "We've got to do something about this, and enact a bill sharing the costs."

Q:

It might interest you to know that one of the new Ontario regulations dealing with acid rain has a provision requiring newer modified boilers. In addition to meeting the ambient air quality standards and in addition to having sulfur content in fuel limited, the deposition of more than .1 kilogram per hectare per year of wet sulfate, as defined by a model which is named in the regulation, is prohibited. This may be exactly what you were suggesting Congress might be looking for.

A: (Wooley)

In New York we used a model to decide what level of deposition we had to establish in order to protect our resources. We tried to figure out how much of the deposition is New York's responsibility using various model calculations. We are in the process of meeting those goals. Eventually it will be a sizeable reduction in New York's emissions. In 1984 Ruckelshaus had EPA go through the same process and they got the same answers but we don't know what they are. The word came down, "No we're not going to do anything about it."

Q: (Moderator)

Would you advocate that Congress choose one particular long-range transport model knowing the adequacy of those available?

A: (Wooley)

No, because I believe we've got to get a bill that leaves EPA no discretion at all to play around with a model in any way. I would like to see a figure, an initial reduction figure and give EPA and the states discretion as to how they reach it. I think we need a more direct approach in this country than is used in Canada.

Q:

Mr. Carter, does EPA have the authority to impose a uniform emissions standard for all of the U.S.? For example, I know in parts of New York State, Consolidated Edison is only allowed to burn up to 3/10 percent sulfur fuel, while some locations in the midwest can burn up to 3 percent and even higher without any controls. It does seem to be unfair, and so my question to you is can EPA impose some base emission limitations? Obviously none of these sources in SIPs (State Implementation Plans) cause exceedences or violations of standards -- that's why their SIPs are approved, but they may contribute to an overall sulfur loading downwind. So does EPA have the authority to impose a standard?

A: (Carter)

Certainly not directly, in the way that you are suggesting. I mentioned the distinctions between the Clean Air Act and the Water Act. The Water Act's structure is pretty much what you're suggesting. There we have a two-step technology-based approach to impose an across-the-board limitation on each particular source and type of effluent. In the Water Act, the agency establishes uniform regulations that for a particular industry category apply across the country. For better or worse, that is not the scheme which Congress adopted in 1970 when it wrote the statute, but went instead with this ambient-based approach which results in different emission limits based on industry, population and so on, resulting in the modeling debate. I'm talking about existing sources only, you understand, because with new sources, we have that authority under the New Source Performance Standards of Section 111, which I didn't discuss. That express provision applies only to new or modified sources, not to all of the existing sources.

Q:

Mr. Wooley, one of the problems New York State has is a credibility problem. I know that some of the sources in New York State are allowed to burn higher sulfur fuel than the SIP normally allows -- they're called variances -- and to use basically the same technique as Indiana, Illinois, Ohio, every other state, and say, "Well, sure, localized impacts show no exceedences of standards so we will allow you to burn for an 18-month period, a 2-year period, something higher than what your SIP allows." Now, if I'm a downwind state, a New England state, I might be a little upset with you yelling and screaming about the midwest when you are doing the same things in New York.

A: (Wooley)

Especially Connecticut. Well, those "special limitations" are actually a revision to a SIP that EPA approved. That doesn't really defend them but we mended our fences with Connecticut because we no longer use that approach. A lot of those special limitations have expired and are not being renewed. The state passed the Acid Deposition Plan which will reduce our emissions 30 percent by the early 1990s. There have been other decisions too. Under the current Commissioner of DEC, an issue came up with the Consolidated Edison plants wanting to convert to coal in New York City. Of course Connecticut said, "Wait, you can't do that, we've got enough already." The Commissioner made a very courageous decision. He said, "You can convert to coal, but first you have to install scrubbers. You have to have absolutely the best controls possible." And we've worked with Connecticut on these litigation efforts jointly so I think the New England States and New York have an identity of interests. New York has got a lot better in recent years and we're continuing to work. There's even a program coming up on further reductions of NO_x emissions so I think at the present time we have a pretty good record.

Q:

I have a question for Mr. Wooley. My understanding is that under the current law, the Clean Air Act, sulfur emissions have come down since the mid-1970s, twenty percent or maybe more. Yet the average annual pH of northeast rain has stayed fairly constant, about 4. How does that square with your present position on uncertainty, that we already know enough to require further reductions?

A: (Wooley)

There was another thing that was going on around the time that the emissions were reduced, and that was that we were building tall stacks, taller and taller smokestacks. So my own perspective is that we are getting more long-range transport at the same time that we are getting some reductions. Also, a lot of those reductions occurred in places like New York City which doesn't impact very much on the Adirondacks compared to eastern Ohio. Some of the more recent studies show that there has been a slight reduction in sulfate concentrations in the streams which seems to parallel somewhat the overall reduction in emissions. What we have pointed out strongly in our evidence presentations is the growing consensus that, basically, there is a one-to-one ratio between sulfur dioxide emission regionally and acid rain regionally. And that serves as a fairly reasonable basis on which to develop an acid rain program.

Q:

Certainly we've heard a lot about ozone recently in terms of the acid rain question. One of the questions that I think needs to be addressed is the relative importance of ozone and sulfur-type pollutants. If the pendulum is swinging toward ozone, it seems strange in some ways that New York State would be pointing out of state, when it has not attained (standards) for areas within the state for ozone which possibly impacts its own agricultural resources. So would anyone like to address that issue?

A: (Wooley)

Do you mean the ozone aspects of acid rain, or on its own - the ozone standard?

Q:

The ozone standard <u>and</u> ozone in terms of acid rain. You certainly have problems with ozone within New York State that might be where the state should put its attention rather than going out of state.

A: (Wooley)

In terms of sulfur and nitrogen, we have an ongoing program. In terms of ozone, we're trying to do that as well. Our Department of Environmental Conservation has the responsibility for developing a SIP, responding to EPA's demands, and it's true, like most big metropolitan areas, our downstate region doesn't attain the ozone standard. Our SIP probably is inadequate to do so. All around the country municipalities and states are trying to wrestle with this problem. What do we do next to attain the ozone standard? I think that New York's ultimate response is going to be very much like its action on the acid rain law, which is that we're going to have to commit to do everything we can about our own sources but also we want reductions upwind. In New York's case, the ozone problem is also very much a regional problem, mostly eastern seaboard. We have a lot of regional air pollution problems, and over the next few years it will be very exciting to see how it turns out. California is probably the leader on the ozone problem.

Q:

Inhalable particles, visibility degradation and ozone, none of which are adequately covered by the Clean Air Act, are generally recognized as regional problems. In all three cases when the state implementation programs are devised to respond to exceedences of these pollutants, it will become quite clear that it is beyond the capability of individual states to meet standards within their borders unless regional sources are also taken into account. Like acid rain, the question will arise whether source-receptor relationships are sufficiently well understood to lead to an overall plan to satisfy the standards within each state. The states will say that it is impossible to meet the standards by rolling back emissions within their own borders, and that we need to have a region-wide if not national program for reducing not only nitrogen oxides, sulfur oxides, volatile organic compounds but also whatever contributes to the inhalable particle mass, visibility degradation or ozone levels. What will EPA do when that day arrives?

A: (Carter)

I'm not sure I can give you a hard and fast answer because at this time the one area I'm familiar with is the ozone situation. My impression of the ozone problem is that the agency and the states are looking for ways to get better control of the precursors and do everything that is reasonable to bring about attainment of low levels. California is on the cutting edge of ozone control with the situation in the L.A. basin which has perhaps the worst problem in this country. At this point, with the existing statute, the agency is trying to figure out what it can do beyond the present control measures to address the ozone problem. I'm not sure how advanced our progress is with the inhalable particle situation at this time because we don't even have a standard on the books yet. And with visibility I can't really offer a suggestion at this point other than to hope that you can come up with and identify a good source-receptor relationship and get to the source of the problem.

I don't want to leave a lingering impression that we never have to deal with uncertainty at the agency or that we are waiting for certainty to be achieved -- that is simply not the case. We deal with it all the time. With the program where I'm most closely involved, Section 112, the hazardous air pollutant program, we don't deal with the range of informed judgment and hypothesis to meet a program. We are not able to achieve certainty but we can narrow the range of uncertainty so that we can make reasonable and informed judgments in order to draft regulatory programs where we can deem those programs to be appropriate. That's been the debate for five years. When do you arrive at a point where you judge that you have sufficient knowledge to draft a solution? At this point the agency is still at the point where it doesn't believe it can act on any reasonable basis in this area.

Q:

I'm not sure this will lead to a question, but if you will let me just make an observation. I want to talk to you about a source-receptor relationship that I heard about many years ago. The source was a couple of double-barreled shotguns that had different shot in the ammunition. The receptor was a dead duck that my brother shot at and my cousin shot at. They started a research program to try to figure out what killed the duck. They pulled the duck apart and started counting the shot which were of different sizes to see who had the most shot inside the duck. They finally made the determination that one gun had clearly deposited more BBs inside the duck than the other. The question was raised where were those BBs in the duck, and can we ever establish which BB or BBs killed the duck? The argument is still going on between those two fellows and they haven't decided yet whose gun was responsible. But I say if we want to have less dead ducks and less dead trees and less dead lakes we stop the smoking gun.

A: (Orloff)

Now I think we know where some of the parables come from. This evening we have seen how lawyers conceptualize the problem of acid rain. Tomorrow during the day we are to be treated to how scientists conceptualize the problem of acid rain.

SOURCE–RECEPTOR RELATIONSHIPS: SCIENTIFIC AND TECHNICAL ASPECTS

Welcome

Elizabeth Thorndike
Center for Environmental Information
33 South Washington Street
Rochester, NY 14608

I want to welcome you today on behalf of the Center for Environmental Information and also on behalf of the 20 co-sponsors of this meeting. They represent the principal constituencies and disciplines of both the U.S. and Canada concerned with the acid rain source-receptor issue.

I would also like to acknowledge our appreciation to the Electric Power Research Institute, the Gas Research Institute, and the National Acid Precipitation Assessment Program, for their support in helping to underwrite the costs of the conference.

One of our long-time practitioners in the environmental arena recently noted that we can now measure parts per billion of chemicals in groundwater, but we can't measure the value of a child's day in a park or wilderness area. That observation gets to the very nub of the problem that concerns us today.

This conference offers you an opportunity for better understanding of the scientific and legal complexities of the source-receptor relationship. However, public perceptions, not facts seen by experts, determine the outcome of public policy issues. The acid rain problem will not be resolved on the basis of scientific or legal certainties but by the value judgments of the decision makers who are involved. With that observation I would now like to turn the program over to the moderator for today, the unanimous choice of the organizing committee, Dr. J. Christopher Bernabo.

Introduction

J. Christopher Bernabo
Science and Policy Associates, Inc.
1350 New York Avenue, NW, Suite 400
Washington, DC 20005

There are lots of acid rain meetings. You, who have been involved in the genesis of these ideas, know that we don't need more sessions where modelers argue with each other about how the systems can be made better.

What we need is a dialogue which transcends the specifics of models or policy making, taking a look at the needs of users of data derived by the modelers and by other available tools.

What are these tools? What are the uncertainties involved in their design and use? What are their applications, uses and values?

How can we produce the proper data unless we know how and for what purpose it will be used?

So our objective here is to bring together users and producers and encourage people on both sides to work out their ideas and proceed.

Our hope is to have presentations here which will in the most general sense explain how these things come together. I think that if you look at the millions of dollars that are spent developing models and the thousands of hours that are spent making policy, you'll be struck by the fact of how little time is spent examining how these two activities can be more productively related. That's really what this conference is all about.

Why We Need to Understand Source–Receptor Relationships

Ralph M. Perhac
Environmental Science Department, Electric Power Research Institute
Box 10412
Palo Alto, CA 94303

A somewhat formal and technical definition of the source-receptor relation can be put in the form of the question, "What is the quantitative relation between emissions and receptor exposure?" Or, using the vernacular, "Who does what to whom?" We know that many industrial emissions contain harmful substances and that society is concerned with alleviating the damage that may result from such emissions. It is in addressing this concern that source-receptor relations play a role.

The source-receptor relation has two important aspects, a geographic one and a quantitative aspect. We tend to consider the two as being inseparable but they need not be. To some extent, the two can be mutually exclusive.

A very obvious application of source-receptor relations is the protection of certain specific regions. We hear, for example, of the need to reduce acidic deposition for protection of the environment. When we hear that need expressed, perhaps we should ask, "Where should deposition be reduced?" If we wish to reduce deposition everywhere, we could simply reduce emissions from as many sources as possible over the entire region. But such widespread, blanket emissions reduction may not be necessary nor the most economic approach to environmental protection. It may be much more practical to focus attention on reducing deposition only in certain so-called sensitive areas. If society opts for selective deposition reduction, then knowledge of the source-receptor relation is essential for designing an effective emissions control strategy.

Some may argue that reducing deposition only in certain selected areas is not a wise environmental policy. They argue that air quality affects a wide number of environmental issues: acidic deposition, human health, visibility, just to name a few. They will further argue that some of these environmental concerns are not restricted to specific "sensitive" regions, that they affect all parts of our nation, hence our strategy should be to reduce emissions from as many sources as is practical without regard to geographic distribution. It would seem, therefore, that adopting a blanket emissions reduction program precludes source-receptor information. It does not. The source-receptor relation plays a key role in quantitative considerations. If we call for a certain reduction in deposition or atmospheric loading, the source-receptor relation tells us how much of an emissions reduction is needed to achieve the desired goal. Or if we call for a specific amount of reduction in emissions, the source-receptor relation can tell us how much of a reduction in receptor exposure we can expect. This is particularly important when secondary pollutants are involved, because the chemical transformations include multiple pollutants in nonlinear reaction systems.

Acid Rain: The Relationship between Sources and Receptors
James C. White, Editor

If we have any interest at all either in the spatial aspect of control or in the quantitative relations between emissions and accumulation, we will want as much information as possible on the source-receptor relation. I am tempted to say that, in regard to air quality, the source-receptor relation will play an important role whatever approach society takes towards environmental protection.

Atmospheric Chemistry — a Lay Person's Introduction

Jeremy M. Hales
Atmospheric Sciences Department, Battelle Northwest Laboratories
Richland, WA 99352

This is not going to be an in-depth presentation of the state-of-the-art in atmospheric chemistry; it will, rather, be targeted primarily for the policy decision community with the hope that the interface between atmospheric scientists and this community can be made more effective.

I shall attempt to give a sufficient overview of atmospheric chemistry so that you can get a firm grasp on just what the key elements are, what the state-of-the-art is, what are the primary areas of uncertainty and how they affect our ability to predict the source-receptor relationship.

I will focus on macroscopic features of the source-receptor relationship. The gentlemen who will speak later in this program are going to be looking more at the microscopic areas. I emphasize that the microscopic phenomena occurring during the source-receptor sequence are very important determinants of just how these systems work.

I'm going to take a very liberal interpretation of the definition of atmospheric chemistry because atmospheric chemistry is a strong determinant of the source-receptor relationship. Within atmospheric chemistry, I'm going to consider as fair game all aspects of "real" chemistry, that is, heterogeneous air chemistry, aqueous phase chemistry and homogeneous air chemistry. In addition, however, I'm going to relate these to wet deposition and dry deposition and give you some ideas on the key uncertainties in both of these areas.

Heterogeneous air chemistry is air chemistry that occurs on the surface of aerosol particles or other environmental surfaces. It involves agents which absorb on the surfaces, sit there for awhile and then react. These products may remain fixed after they react, or they may return back to the gaseous state.

Aqueous phase chemistry is chemistry that can occur within raindrops, within cloud droplets, even within the ice phase of snow crystals.

Homogeneous air chemistry is totally gas phase chemistry and is the aspect of atmospheric chemistry where our understanding is the most well developed at this time.

Wet deposition and dry deposition tend to be aggregate products of all of these chemical processes and I'll be leading into these in the latter part of my talk.

Our present goals thus can be summarized as follows. First to provide an intelligent layman's overview of air chemistry and deposition properties with emphasis on policy analysis and evaluation. Second, to demonstrate why these processes are important for source-receptor evaluation and to try to answer

Transcribed from remarks recorded at the conference.

Published 1988 by Elsevier Science Publishing Company, Inc.
Acid Rain: The Relationship between Sources and Receptors
James C. White, Editor

the questions, "Who cares?" or, if you don't care, "Why should you care?" Finally I'm going to indicate the present state of understanding in these fields and I'm going to identify the key issues affecting our concern in these areas.

I'd like to begin by adding some support to the comments of Ralph Perhac on why atmospheric chemistry is important in the source-receptor relationship. First, atmospheric chemistry affects the amount and chemical state of the deposition material and secondly, it affects the region of deposition.

In Figure 1 we have a situation where material A is being released at a point source. A is capable of being reacted to material B if chemical and physical conditions are right. A and B have different properties such as deposition rates and the atmospheric chemistry occurring here will influence their relative concentrations. One sees that unreacted A will probably travel a long way prior to deposition but resultant material B will probably precipitate closer to the point source.

Thus the atmospheric chemistry influences both the <u>nature</u> of the compounds and the <u>region of deposition</u>. This translates into a different source-receptor relationship.

Nitric oxide reacting or not reacting to form nitric acid would follow this pattern as would sulfur dioxide being converted to sulfuric acid.

I shall now zoom in on specific aspects of air chemistry and I'm going to talk about this in several phases, homogeneous air chemistry, wet deposition, dry deposition, the aggregate products.

I'd like to begin my discussion on homogeneous air chemistry by asking you to visualize with me a reacting pollutant molecule that happens to be "buried" in a medium of air. A guiding rule of atmospheric reaction kinetics, or any reaction kinetics, is that two pollutant molecules must collide with each other before they can react. That is a necessary condition for the reaction to take place, but it is not a sufficient condition. If they collide and bounce off of each other, they'd be totally unchanged as a result. Some of those collisions, however, are going to result in a reaction between the species. That poses an interesting question in my mind. Here we have one reactive molecule of a pollutant mixed with perhaps a billion inert air molecules. It's going to be real tough for this molecule to find another pollutant molecule to collide with at these low concentrations. In fact, at this parts per billion concentration there will be a billion or so collisions with air molecules before one occurs with another pollutant molecule. The big question is, in this sea of air molecules, how do these pollutant molecules ever find each other in order to react?

It is amazing that not only do they find each other to react but there is an intricate, interlocking web of chemical phenomena that tends to chain the reactions of these molecules together.

One major reason for this chemical reactivity is that there are lots of collisions between molecules in the atmosphere. If

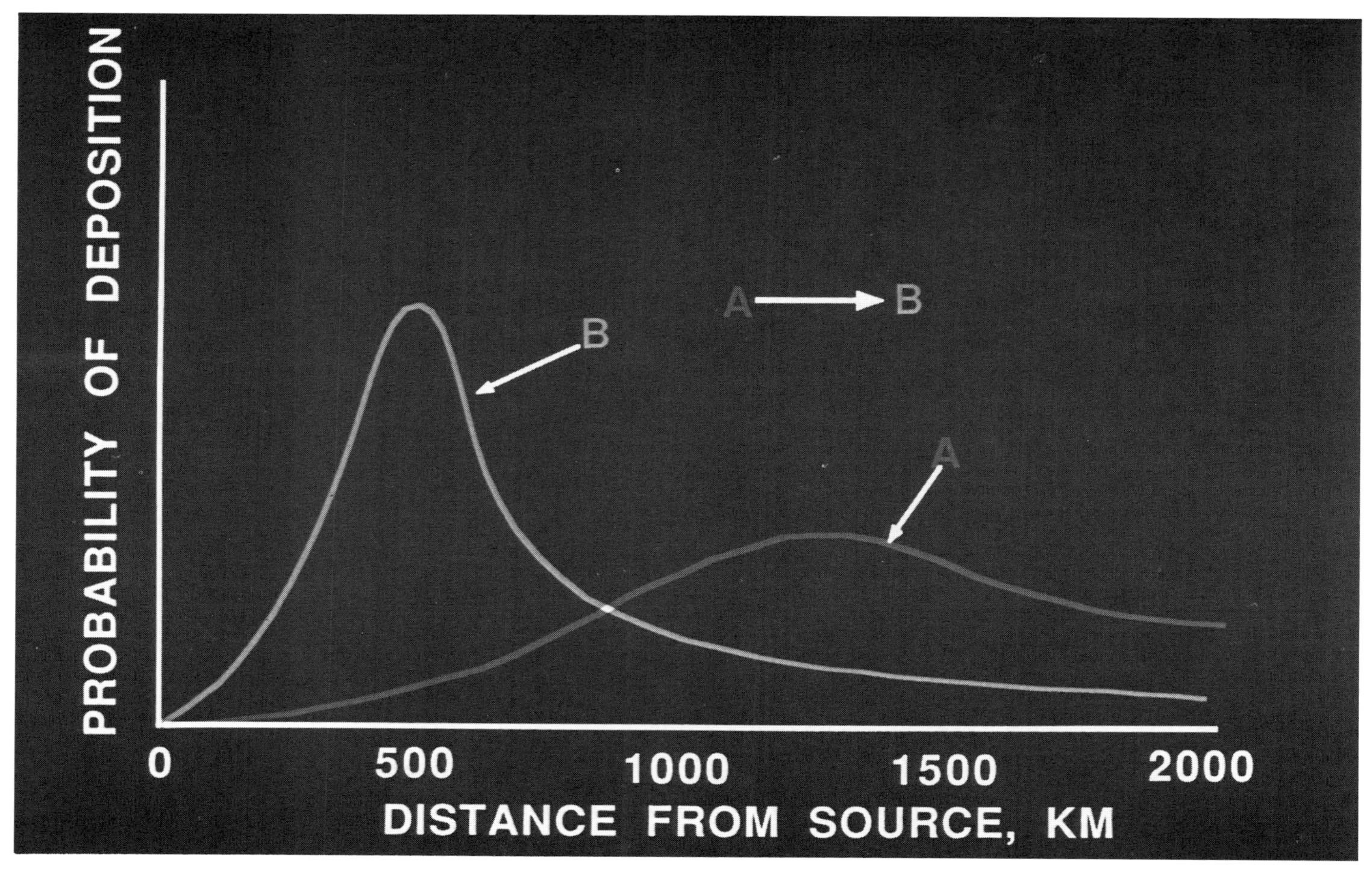
PROBABILITY OF DEPOSITION
A → B
B
A
0
500
1000
1500
2000
DISTANCE FROM SOURCE, KM

Figure 1.

you were to calculate them it would be something between 20 million and 100 million collisions of a given molecule with its neighbors per second. Some pollutants are very reactive, and when they collide the probability of a critical chemical reaction is very high. Also, and very importantly, the sun serves to provide a source of photons which tend to activate certain chemical species to excited states. These excited states split to form fragments of molecules also in excited states, and those can go on to react with other species in a very effective fashion.

There are a limited number of reactions that can absorb photons from the sun and troposphere to get this type of chemical interaction going and I'm going to refer to these here as the "trigger" mechanisms.

If we plot (Figure 2) light intensity as a function of wavelength from the ultraviolet to the visible spectrum and into the infrared, we observe that the solar spectrum near the lower atmosphere does not proceed far into the ultraviolet region. Ultraviolet is the more energetic fraction of solar radiation and is the agent which tends to form most of the photochemically excited species. There are several reactions that occur within the lower troposphere, within the visible wavebands, that are very important trigger mechanisms for photochemical reactions. One of the very important ones is the photodissociation of nitrogen dioxide gas which occurs over a fairly broad range of the near ultraviolet visible spectrum from 290-430 millimicrons to form nitric oxide plus oxygen atoms. Oxygen atoms tend to be very reactive and they are essential in setting off some of the changes that occur in the subsequent chemical reactions.

Ozone itself can photolyze to form oxygen atoms. Nitrous acid, HONO, also can be photolyzed, this time to form nitric oxide plus a hydroxyl radical. I'll come back frequently to hydroxyl radicals during the talk. Hydrogen peroxide can photolyze to form a couple of hydroxyl radicals. These are certainly not all of the tropospheric photolysis reactions but they're some of the important few. Its not particularly important for you to remember these reactions exactly. It is important, however, that you gain the visual image that photochemical triggering processes exist that are very important determinants of atmospheric chemistry.

There are also central features in the atmospheric chemistry sequence that tend to wed the pollutant chemistry of sulfur species, carbon species, nitrogen species, even halogen species together. To illustrate this I'm going to apply an allegorical reference first used by Dan Albritton of NOAA where he depicted ozone as the pollutant "godfather", and hydroxyl radical as the "hitman". As noted previously, ozone photolyzes within the visible and near ultraviolet forming oxygen atoms which react subsequently with other species. It's important to remember also that ozone can react with nitric oxide to form NO_2. That's important because NO_2, as I indicated before, is a very effective photolytic source of reactive oxygen atoms. So ozone plays an essential role in setting the stage for all of the other chemistry that goes on behind it. And one of the reasons it plays this essential role is that it sets the stage for the "hitman". The oxygen atoms shed by the photodecomposition of ozone are reactive with water vapor

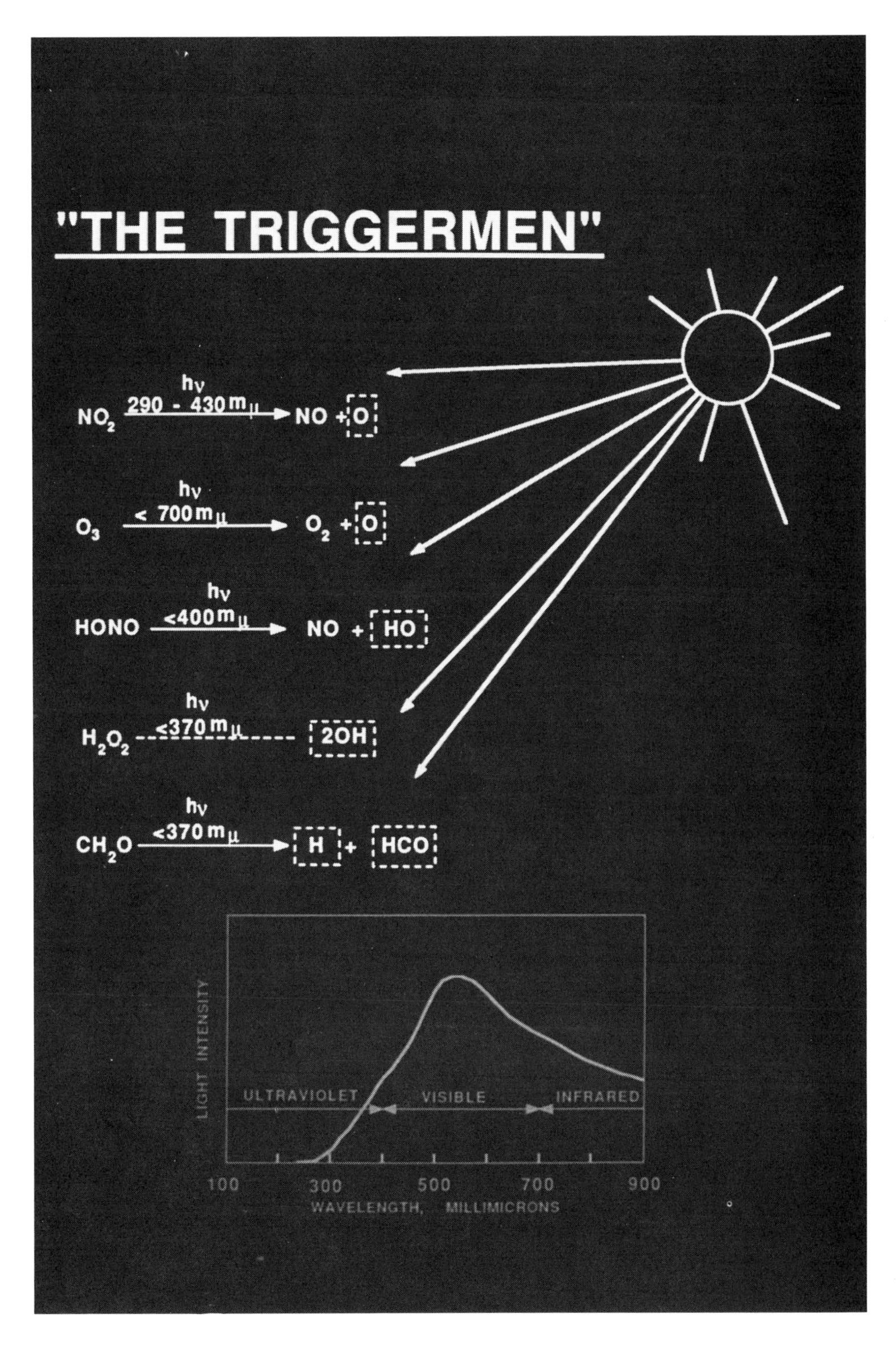

Figure 2. Light intensity as a function of wavelength, and reactions triggered photochemically.

molecules to form hydroxyl radicals. This oxygen also can react with hydrocarbons again to form hydroxyl radicals and organic free radicals. In addition, atomic oxygen can react with H_2O_2 to form hydroxyl radicals and also HO_2. These are not the only mechanisms for formation of hydroxyl radicals but they are probably among the most important.

The hydroxyl radical is a tremendously effective oxidation agent for converting SO_2 to sulfate or sulfuric acid in the atmosphere. It's a complicated reaction and there are some aspects of it that aren't totally resolved at this time. But we do know that it is technically a very important reaction. Hydroxyl radicals also react with nitrogen oxide to form nitrous acid and with NO_2 to form nitric acid. Both the sulfur and the nitrogen pollutants which are at the cornerstone of importance in acid rain and the source-receptor relationship are formed by the action of the hydroxyl radicals. Again it is not extremely important that you remember these chemical reactions but it is important that you maintain the three mental images of ozone as the "godfather," hydroxyl as the "hitman," and the "trigger" reactions of the sun.

Today we attempt to include atmospheric chemical reactions such as those in Figure 3 into source-receptor models, but this is not an easy task. The mechanics of this modeling process require that we be able to express mathematically the reaction rates and the concentration of reactants. More and more elaborate chemical schemes are being included in these models as our chemical understanding and computer resources expand.

The photochemical parameterization by Friedlander and Seinfeld in 1969 was one of the first attempts to describe the complex of pollutant reactions in a mathematical sense. They set forth rate coefficients and rate equations for the photochemical smog process with the following equations:

$$NO_2 \longrightarrow NO + O$$

$$O + O_2 \longrightarrow O_3$$

$$O_3 + NO \longrightarrow NO_2 + O_2$$

$$\text{Hydrocarbon} + O \longrightarrow \text{Free Radicals} + \text{Products}$$

$$\text{Hydrocarbon} + O_3 \longrightarrow \text{Free Radicals} + \text{Products}$$

$$NO_2 + \text{Free Radicals} \longrightarrow \text{Products (including PAN)}$$

This early effort represents a turning point in atmospheric modeling.

I intended to provide you with examples of some of the more modern reaction schemes but I find it impossible to depict these here in an understandable manner, owing to the multiplicity of reactants. Some of these schemes pose a real computational challenge, even for today's supercomputers.

We should itemize a few of the currently pressing issues in homogeneous air chemistry. One of these is accounting for all the nitrogen oxides. There is a lot of it floating around. It's

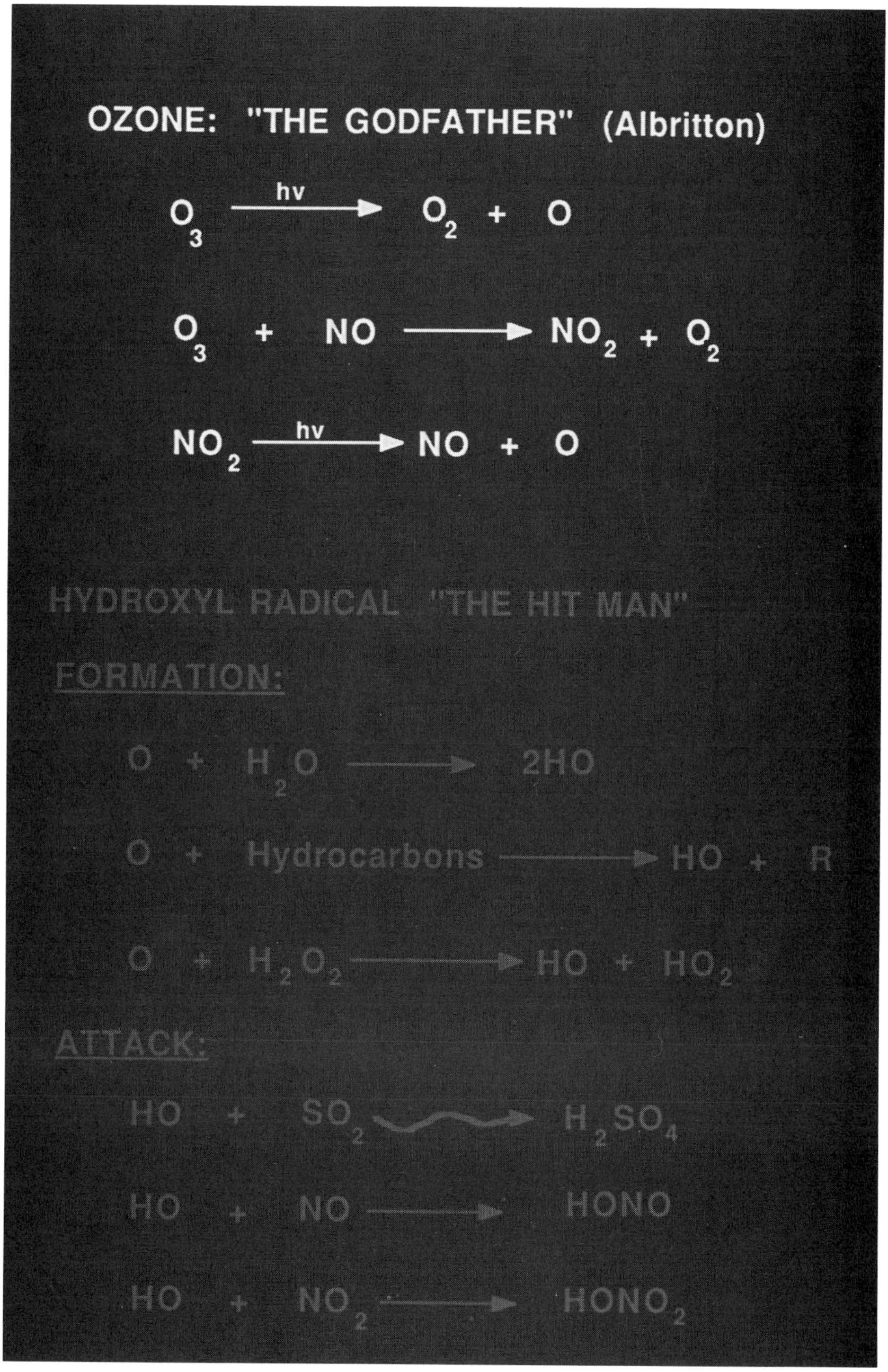

OZONE: "THE GODFATHER" (Albritton)

$$O_3 \xrightarrow{h\nu} O_2 + O$$

$$O_3 + NO \longrightarrow NO_2 + O_2$$

$$NO_2 \xrightarrow{h\nu} NO + O$$

HYDROXYL RADICAL "THE HIT MAN"

FORMATION:

$$O + H_2O \longrightarrow 2HO$$

$$O + \text{Hydrocarbons} \longrightarrow HO + R$$

$$O + H_2O_2 \longrightarrow HO + HO_2$$

ATTACK:

$$HO + SO_2 \rightsquigarrow H_2SO_4$$

$$HO + NO \longrightarrow HONO$$

$$HO + NO_2 \longrightarrow HONO_2$$

Figure 3. Reactions of ozone and the hydroxyl radical with air pollutants.

an uncomfortable fact of life that we cannot balance all of the nitrogen oxide that is going in and out of the different chemical states in the atmosphere. There are some species that we just don't know about that contain oxides of nitrogen and they seem to be involved in reversible reactions. We are not depicting these well in the models and we are not measuring them well in the outside atmosphere. We need to close the gap on this particular area of atmospheric chemistry before we can improve our understanding of nitrogen oxides appreciably.

A second issue is radical measurement. The sad fact is that, even though radicals are super-important in piecing together our understanding of the atmospheric source-receptor atmospheric chemistry process, we can't seem to measure them satisfactorily. We've been frustrated by this for the past 20 years. We know they're out there, but we have poor information concerning their concentrations. These concentrations are very low and this plus their high reactivity make them particularly hard to quantify. We need to know more.

A third pressing issue is knowledge of the role of hydrocarbons. I haven't said a lot about these chemical species, but they are extremely important and their roles are still poorly understood. Because there are so many hydrocarbons, such a wide variety of hydrocarbon mixes and so many reactions, we have to parameterize their effects. We have little quantitative data and lack the ability to predict their aggregate functions well.

The next topic is dry deposition. As an introduction I need to give you a very broad-brush description of what the dry-deposition velocity concept is. We have a source-receptor relationship with material coming out of a stack coming over and impinging on a receptor. We quantify deposition mathematically by the use of the term V_d, standing for deposition velocity and we measure the quantity of deposition by introducing a term for concentration of the material at the receptor site. Thus the expression:

$$V_d \; C = \text{dry deposition flux.}$$

Historically this has been viewed as an empirical relationship, employed mainly because we usually don't have any better alternative. It is common, however, to view this deposition velocity concept in terms of an electrical resistance analog, where the deposition velocity is expressed as a function of a reciprocal of an "atmospheric resistance" to deposition. Figure 4 is a somewhat naive depiction of this analogy which has been used by atmospheric scientists for some time.

It's an implicit assumption of the original dry-deposition velocity concept that the surface is an infinite sink; once the deposition molecule contacts the surface, it does not come back off, and therefore the air-pollutant concentration adjacent to the ground is zero. A somewhat more realistic application of the resistance analogy, however, is to state that the atmosphere is not the only medium that resists deposition; that is, "surface" resistance to deposition exists as well. From this we can extend our analogy to include a resistance for the deposition surface, the ground, a leaf, a concrete bridge

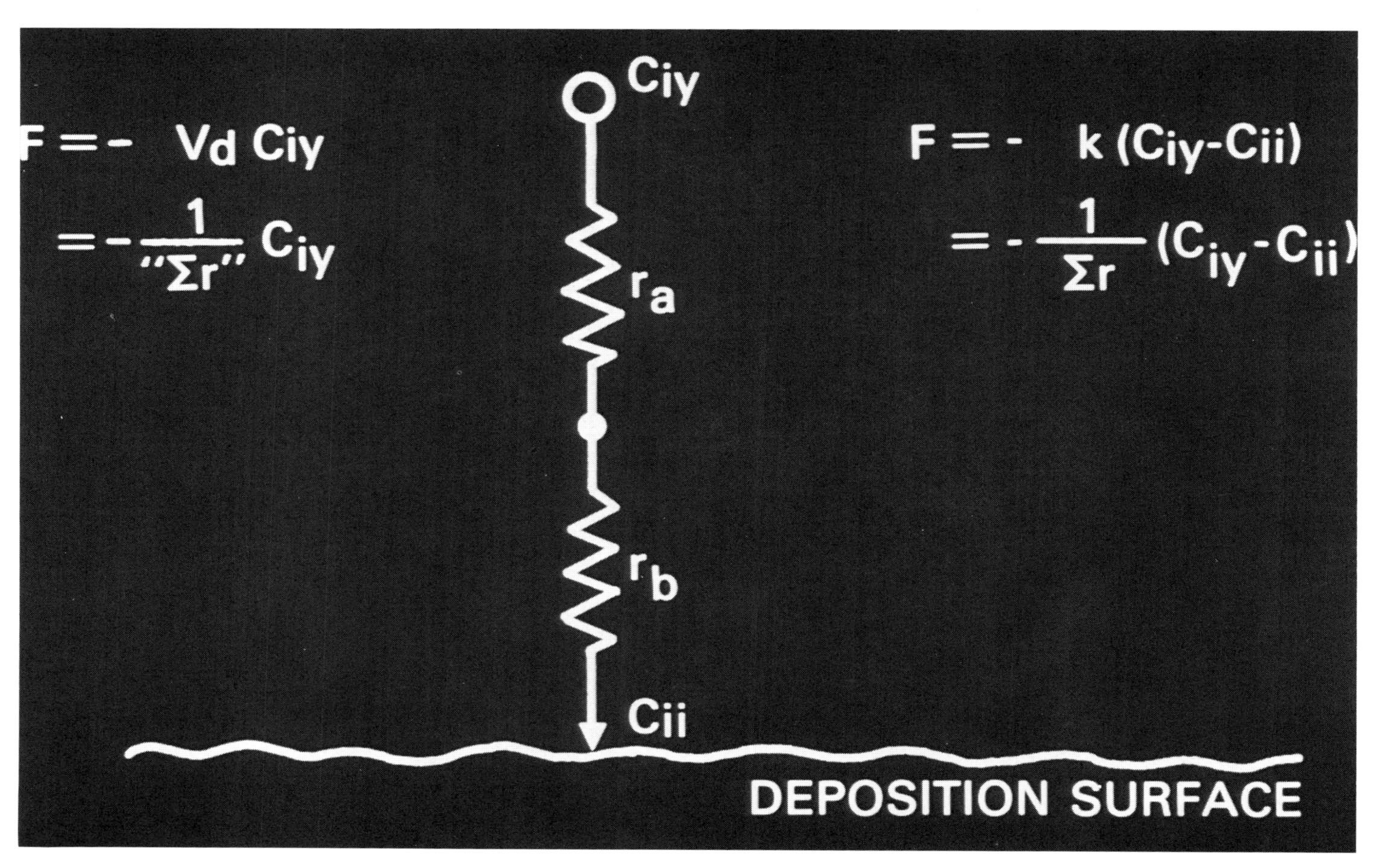

Figure 4. The deposition velocity concept depicted as an analogy to electrical resistance.

abutment or whatever; the important thing is that this surface itself presents resistance to deposition. We depicted this by r_c in Figure 5.

At least some receptors, however, are known to re-release pollutants back into the atmosphere and the time scale of picking it up and re-releasing it from the surface may have marked differences. As a result it has been suggested that, besides "resistance," some sort of "capacitances" should appear in the electrical analogy. This is depicted in Figure 6, where we add a capacitance to absorb, and then perhaps re-release material.

Presently it's becoming more and more accepted that the atmospheric side of the deposition sequence is usually rather simple compared to the receptor side. There are so many processes that go on within the receptor. Just think of the life processes that go on within a leaf. I give you Figure 7, a preposterous representation, to give you a mental image of the probable complexity of both the atmospheric side and the surface receptor side of deposition.

Let us now turn to wet deposition. In discussing this I'd like you to visualize a cloud as an air-borne, aqueous-phase "medium" for a diverse set of chemical reactions.

Let us look at a few of the mechanisms that go on, both physical and chemical, to provide a feeling for the complexity and the importance of cloud chemistry. First let's look at a single droplet of water within this cloud and observe how the droplet processes are affected by chemical reactions. The first thing that's important is the formation of the droplet itself. A pollutant crystal can act as an absorption nucleus to attract water molecules and form a cloud droplet. It's very important that we have these pollutant particles up there because clouds really can't form without these condensation sites. There are lots of nuclei in the atmosphere, however, and even if we took out all the human-produced particles there would be plenty of natural ones left to allow cloud formation.

Water droplets and water vapor molecules can adhere to these nuclei and form cloud droplets but this reversible process can return the water to its vapor state, depending on ambient humidity conditions.

Once the droplet is formed many transport mechanisms can bring gaseous molecules, aerosol particles and pollutant molecules to the surface of the droplet. Among these are diffusion, coagulation, condensation, and various electrostatic and thermal effects. These effects are also reversible. Figure 8 is a simplified presentation of the way water components in the atmosphere interact continuously and reversibly.

Once the cloud droplets are formed, we can go through a series of growth processes that convert the small cloud droplets into larger drops that can settle out because of their weight and size. Coagulation of two cloud droplets can occur to form a larger single droplet. The larger droplet has more tendency to settle out of the cloud. It can grow bigger and bigger through this process until it gets to sufficient velocity to impact additional cloud droplets and form precipitation.

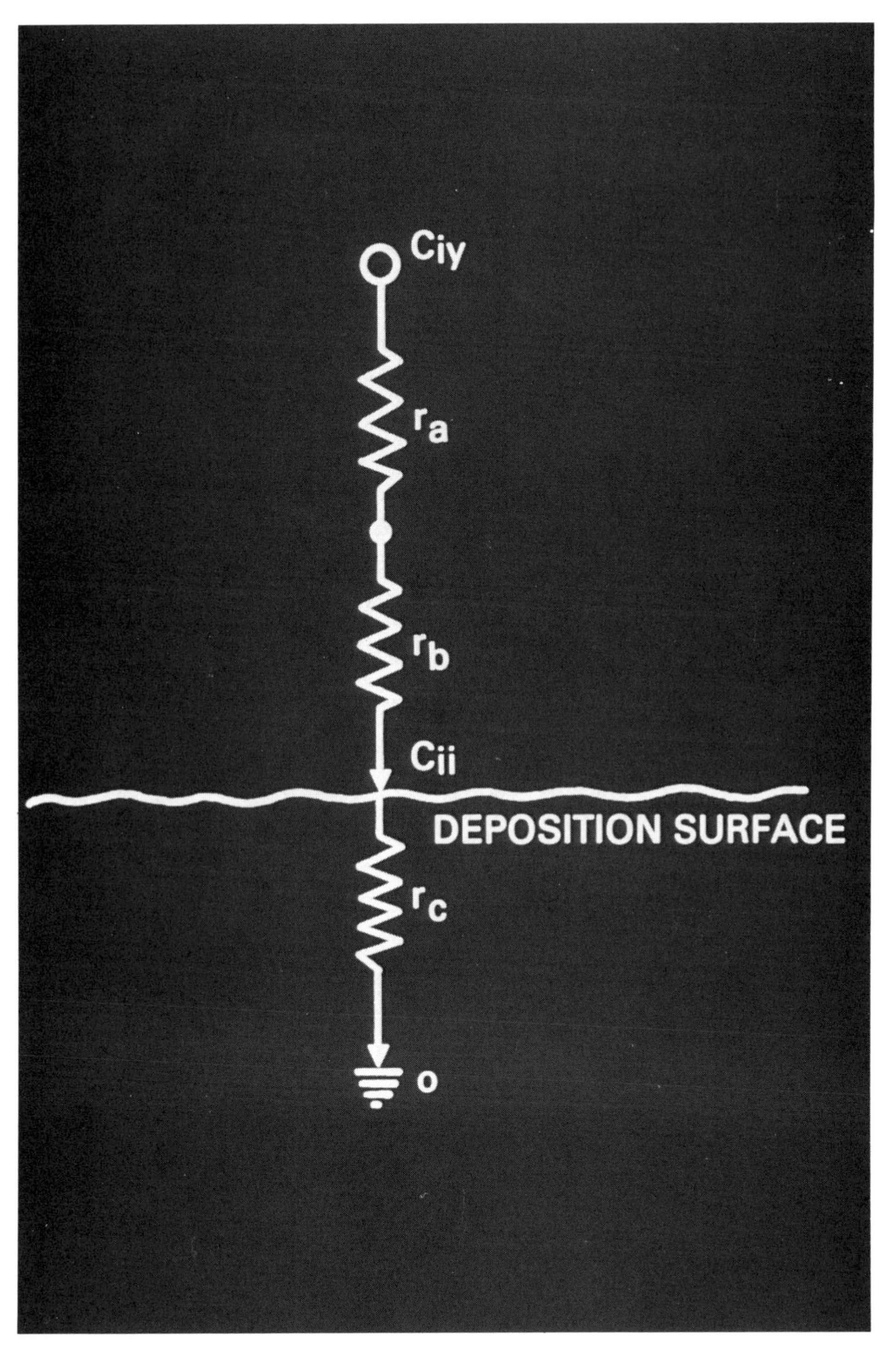

Figure 5. Depiction of surface resistance to deposition.

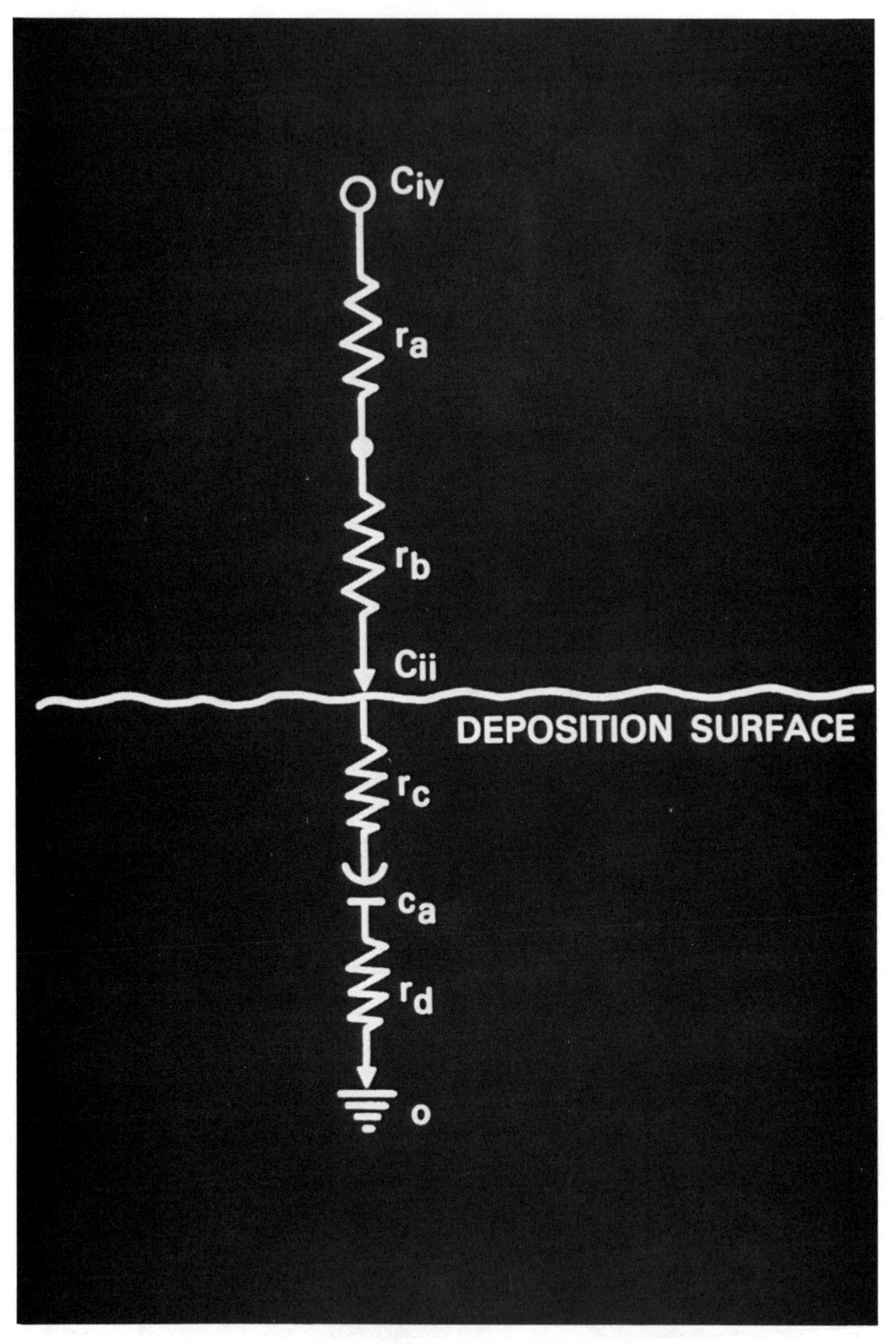

Figure 6. Simulation of surface reaction as a function of capacitance.

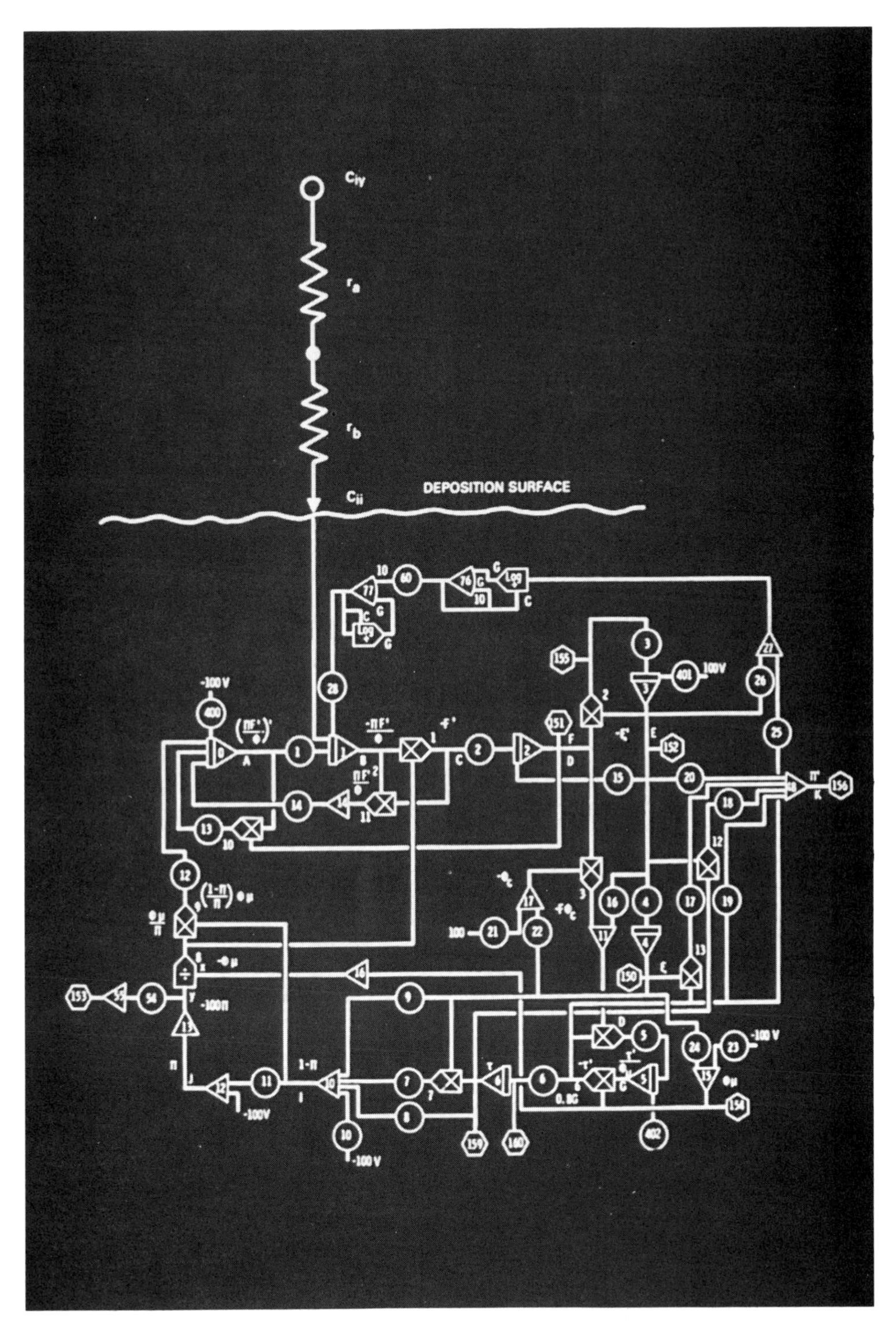

Figure 7. Preposterous depiction of possible complexity of the atmospheric and surface receptor sides of deposition.

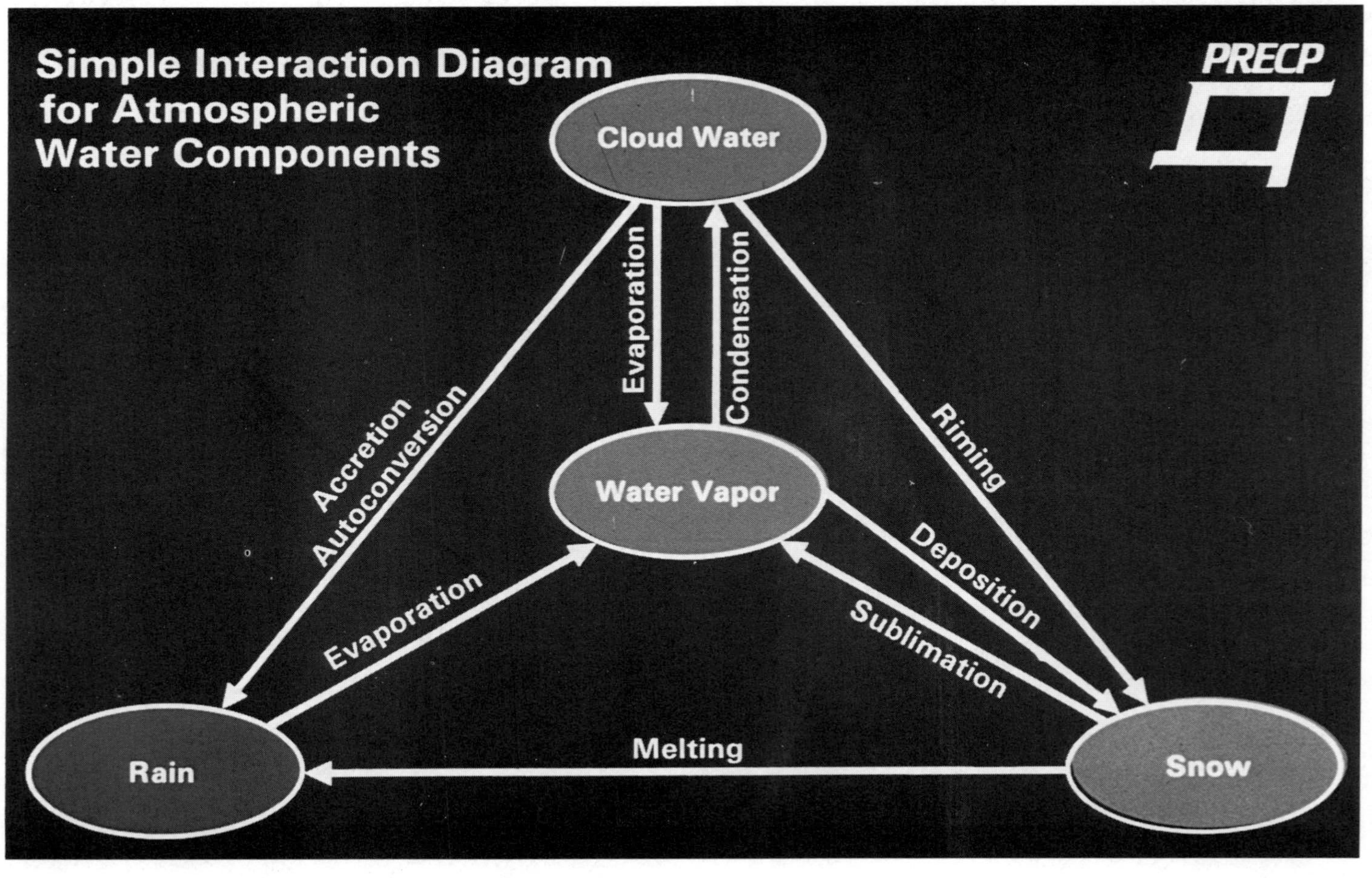
Simple Interaction Diagram
for Atmospheric
Water Components
PRECP
Cloud Water
Evaporation
Condensation
Water Vapor
Accretion
Autoconversion
Riming
Deposition
Sublimation
Evaporation
Rain
Melting
Snow

Figure 8.

We also can have sub-freezing conditions where an ice phase is present. There's all kinds of complicated competition between the ice phase and the aqueous phase, a lot of microphysics.

I'd like to step away now from the pure physical process and go into the chemical process of the aqueous phase. As I mentioned before the chemical processes that are going on within the aqueous phase are not anywhere nearly as well understood as the homogeneous air chemistry processes.

There are some things that we know for sure. We know, for example, hydrogen peroxide is an extremely important oxidant of SO_2 in the aqueous phase. We know also that ozone can be important only under some conditions in non-acidic environments. But ozone oxidation can occur in most acidic precipitation-forming clouds. Clouds tend to be very heterogeneous in makeup and may undergo reactions which release acidity to a point where ozone activity is no longer inhibited.

There are lots of unanswered questions about what is going on within the aqueous phase environment of clouds. What is the role of free radicals? We know that they are active but we certainly haven't worked out the chemistry in any satisfactory form. What about catalysts? We've known for a long time that trace metals and other types of catalysts can enhance the SO_2 oxidation process. When researchers try to measure these in the laboratory, however, there are so many variables that it reveals a very murky field. What about inhibitors? Things that perhaps sequester SO_2 and put it in a form that is not reactive. That's a big item. We know that formaldehyde is an important source for a complex that essentially protects SO_2 from attack by H_2O_2, and we know that there is sufficient formaldehyde around to be effective, but there are key elements of this process that we still don't understand.

What about NO_x chemistry and the aqueous phase? Much is known about aqueous NO_x chemistry but there are also many things that are poorly understood. Present wisdom has it that most of the NO_x oxidation chemistry goes on outside of the aqueous phase with the formation of nitrate aerosols and nitrogen acids. Is this responsible for all of the nitrate that shows up in the aqueous phase? We're not totally sure. There's some data, especially field data, that indicates that this may not be the case under all circumstances.

What about ice-phase reactions? It's been thought, up until the present time, that when you freeze a water droplet, you essentially turn the chemistry off. It turns out that current thinking is wrong. We have late-breaking evidence that there's a lot of chemistry that can go on, maybe not within the medium of the ice phase, but on the surfaces of the ice phases within a cloud system. We have some recent results, for example, which indicate that H_2O_2 and SO_2 undergo a very fast surface reaction on ice particles. It appears that we're going to have to completely rethink this area.

In summary, we have good knowledge of some aspects of aqueous phase chemistry. There are some things that we question, and some things that we just flat don't know. We do know, however, that aqueous phase chemistry is very important for a lot of the reactions that enter into the source-receptor sequence.

The modeling process of assemblying this microscopic physical and chemical information to provide a mathematical representation of wet removal poses a difficult problem.

In early modeling efforts we have visualized the storm as a "black box," attributed some sort of an empirical scavenging coefficient to that box and proceeded with estimates. This is not acceptable for defining the source-receptor sequence, because many mathematical nonlinearities are intrinsic within a storm system.

How do we handle this in more modern modeling practices? One way is to apply what is known as the "interaction diagram" concept. We formulate interaction diagrams that define the transitions from one phase to another and one chemical species to another, or classes of phases or species. Then we set up mathematical formulas that depict those features mathematically.

Let us consider the output from such a model, depicting scavenging by a convective storm system where the top of the ice layer is at 12 or 13 kilometers (Figure 9). A lot of precipitation is falling within the storm mostly as ice, although water is present in the cloud. This output represents a two-dimensional slab through a storm system, and this first figure shows only the water components. Height is plotted on the ordinate and distance along the storm on the abscissa. Note that it's a fairly pronounced convective storm, the top of the ice layer is getting up to the order of 12 or 13 kilometers. The ice, which has a fall velocity, is depicted by the slanted lines. Cloud water shows up as the dots and finally the rain appears at the bottom.

Figure 10 is the same system showing the sulfur components of the storm system. It's a rather messy figure showing stream functions and SO_2 and sulfate isopleths. The SO_2 and sulfate are being blown around by the flows within the storm. SO_2 is reacting within the storm in the aqueous phase, with hydrogen peroxide. Ozone is also acting as an oxidant, but this process is inhibited by acidity until the storm has been sufficiently cleaned for the ozone mechanism to take over.

The solid lines in this figure depict the stream function, showing the flows through the storm. The central portion, where we have these lines grouped together, is a strong updraft region within the storm system. You can see downdraft regions on both the left-hand side and the right-hand side. Primarily, however, this is an updraft region where the SO_2 and the sulfate are drawn up into the storm system. Some of it is being reacted upon, some of it is scavenged, the rest of it is being delivered out the top.

The important thing to note is that the SO_2 and sulfate, which were initially concentrated in the lower portions of the cloud, are being vented extremely rapidly up through the layers and into the upper atmosphere. A lot of SO_2 and sulfate is showing up at 10 kilometers because of the venting activity of storm. This is one thing that we're not depicting well in regional models of precipitation scavenging of pollution.

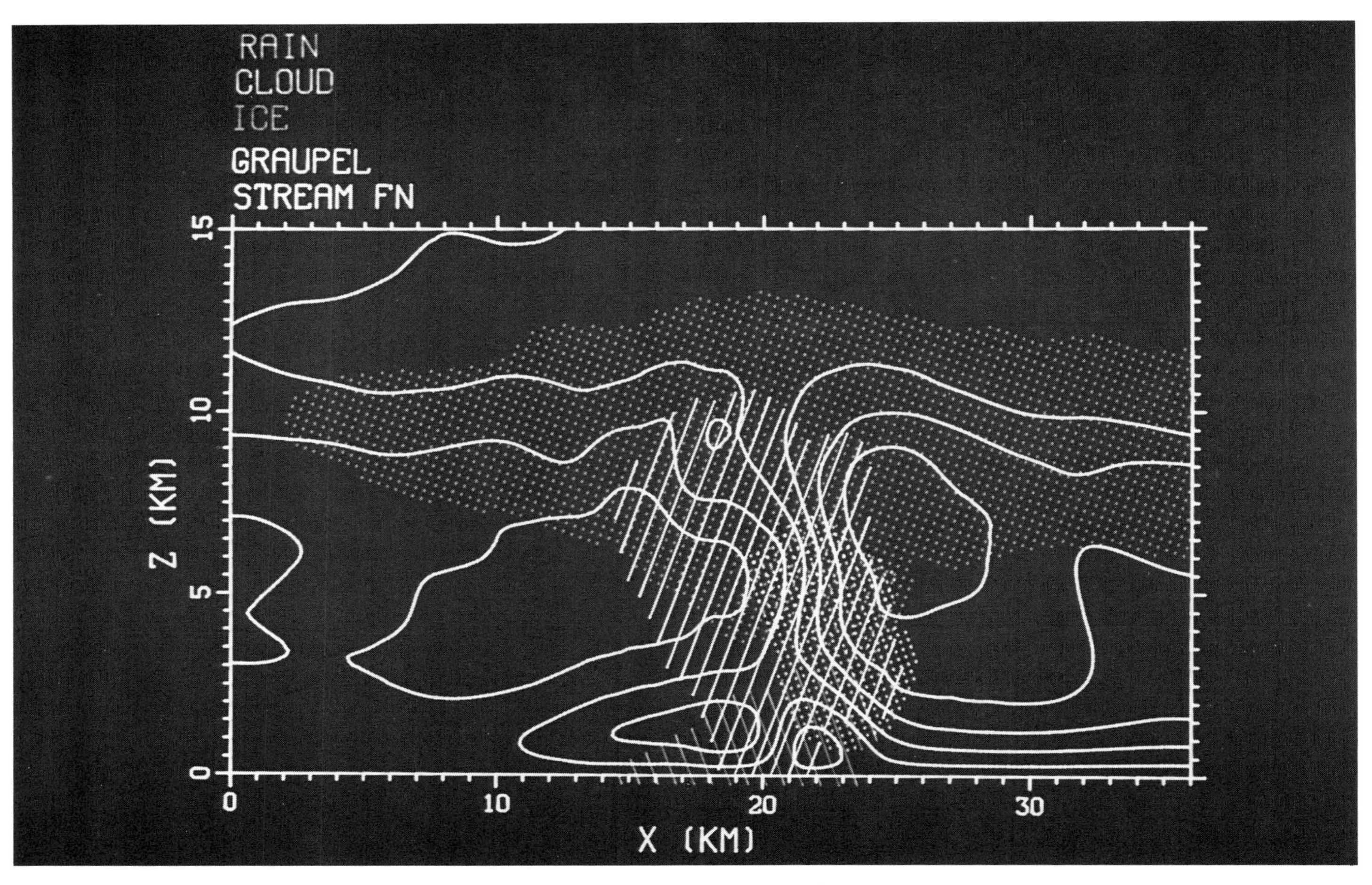

Figure 9. Scavenging by a convective storm system.

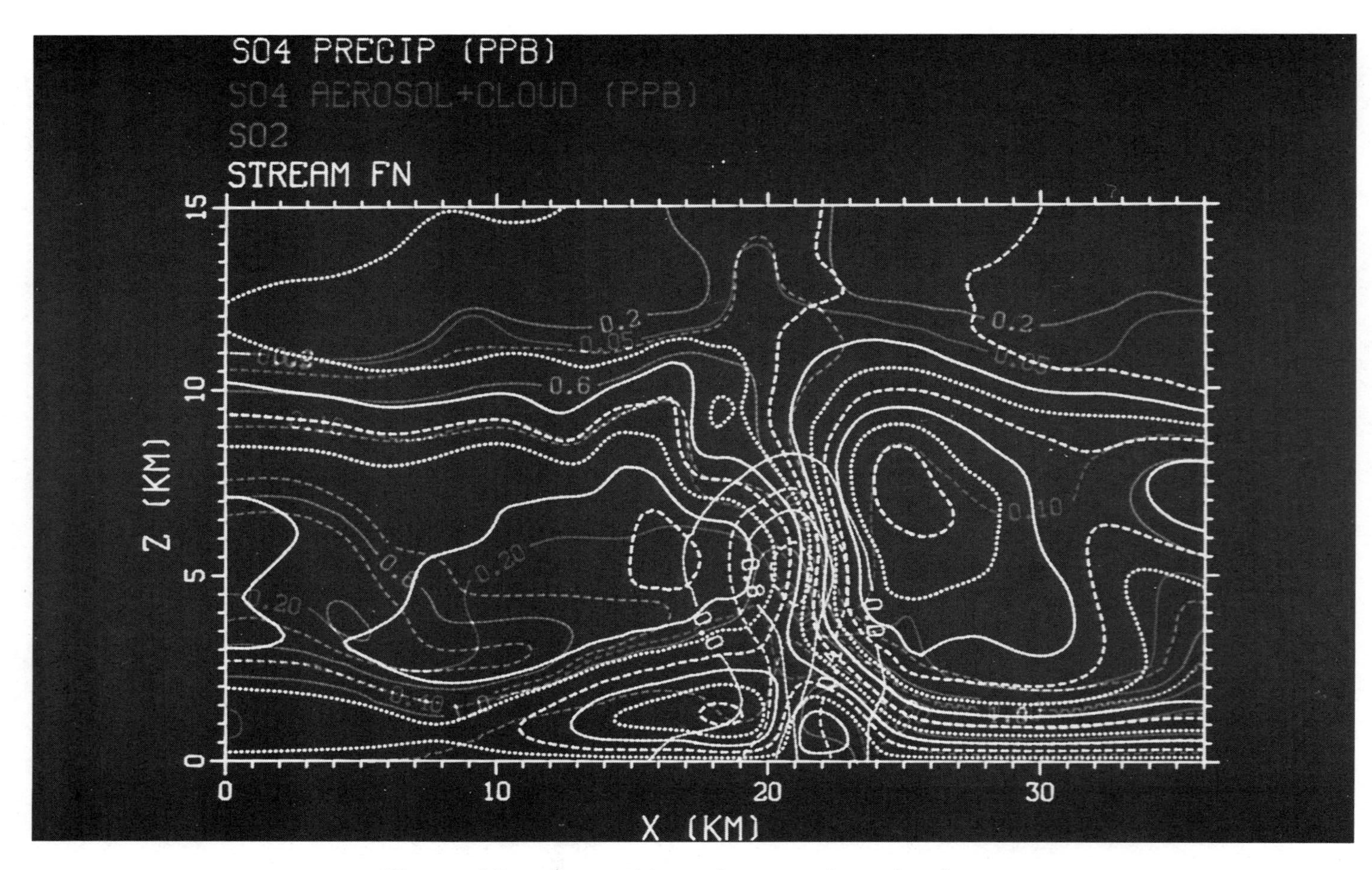

Figure 10. Convective storm system showing strong upwelling.

A big problem that currently exists is that of taking precipitation scavenging phenomena and wet chemistry of this complexity and depicting it in simplified but accurate characterizations so that it can be represented well on large regional models and still fall within computer resource capabilities.

Questions and Answers following Atmospheric Chemistry Session

Thursday, December 4, 1986

Q:

Based on your description of the role of ozone and hydrocarbons in forming the free radicals, do you feel that there is a real possibility of pollution being controlled by Congress in terms of controlling only SO_2 and nitrogen oxides? Might this not fail to achieve the significant environmental impact that we are seeking?

A: (Hales)

There is no question in my mind that hydrocarbons are a key element in the atmospheric gestalt. I'm not quite sure about the policy side of your question, but let me make my own opinion clear that, unless we include hydrocarbons in the formula about the overall control strategy, we are making a mistake.

Q:

I have two questions. One, is it a possibility that pollution may ultimately cause more pollution, like acid rain may kill a forest and then you have trees rotting and adding hydrocarbons to the air? And second, a lot of the chemical reactions seem to depend on ozone, yet we hear talk of freon destroying the ozone layer. How does freon fit into this formula?

A: (Hales)

First, the second question. What I presented for you in the homogeneous air chemistry was a kind of web of chemical reactions that have a very strong flavor of generality over the air chemistry process. A lot of those reactions are important; halocarbons are seen in all layers of the atmosphere, but their bad effects are a stratospheric phenomenon. Here you have a lot more ultraviolet radiation, so all of a sudden you have a lot more reaction. The halocarbons become very important. They're not important in the lower layer.

The first question is really an intriguing one. There are all sorts of feedback mechanisms that could be catastrophic. These have been hypothesized. I don't want to enter into any conjecture about that but, yes, people have thought about it.

Q:

On the question of ozone and hydrogen peroxide, I was under the impression that hydrogen peroxide was a much more important oxidizer for SO_2 than ozone. Would you care to comment?

A: (Hales)

If you look at it from the bench chemist's point of view, where you take a beaker full of SO_2 in water and you add a certain number of moles of hydrogen peroxide, there is no question that the reaction is going to go much faster than if you performed a similar experiment with ozone. There is also no question that the ozone reaction is inhibited by acidity. So, looking at it from that point of view, one would say that hydrogen peroxide must be more important out in the atmosphere. The problem is that cloud systems are a lot more complex than a beaker in a laboratory. We have zones of dirty air which tend to coincide with where the hydrogen peroxide is being exhausted. When it is exhausted, then you have a tremendous volume in a cloud system that is relatively clean, and you have plenty of residual ozone, so just on a volumetric basis, you have more opportunity for reactions. Looking at that argument, and looking at some modeling arguments, you can say that ozone becomes important in that particular phase. As I showed you, ozone accounts for approximately 55 percent of the oxidation, so hydrogen peroxide would account for 45 percent. You can pick a storm where that relationship does not apply, even where the ozone reaction never does get going. I think it is an open question, about which will dominate over the aggregate of the storms that occur in the northeastern United States. I think that current wisdom has it now that hydrogen peroxide is going to be the more important, but you are making a real mistake if you don't include ozone in the formula as well.

Q:

You have shown very convincingly that not all sulfur is treated in the same way in a cloud, and this brings me back to source-receptor relationships. If you had four plants in your model, with each of those four contributing equal amounts of sulfur going into that cloud, because of the nonlinearities in the cloud, I sense that perhaps plant A would contribute more than its share to the total amount of sulfate that came out as precipitation, and your model could demonstrate, perhaps, which of these plants contributed. But how are we going to interpret that in terms of source-receptor relationships if the system is nonlinear?

A: (Hales)

In the general source-receptor relationship, we are up against a wall with modeling, because we need to talk about how the source-receptor relationships occur in a time-averaged, chronological sense. When you take a bunch of storms and run them over the plants, you are going to get a composite. How you do that in a modeling sense with limited computer resources, I don't know. My guess is that with the large regional models which have the processes that I showed highly parameterized, we can capture the essence of the nonlinearity.

Q:

What I was trying to get at was that it is difficult to take the results from complicated models and aggregate them to some sort of long-term statement about source-receptor relationships.

A: (Hales)

I agree with that, and want to emphasize that the model I showed you is a diagnostic model. It's a very complicated model with a lot of physics and chemistry, and is not the same type of model as the regional models that are being developed right now, which have this type of phenomena consolidated and simplified in a substantial fashion.

Q:

Out west in some places you have almost neutral pH clouds, ph 6 and 5.5, and inorganic material, calcium salts and so on, act as a buffering agent. In more acidic areas, you have a SO_2 reaction forming bisulfites and I think the bisulfite reaction is quite important. That may be one reason why acid rain normally has a pH not too much above 4.5.

A: (Hales)

I think so. When I said SO_2 oxidation in the aqueous phase, that was a gross oversimplification, because when SO_2 dissolves in water, it forms bisulfite, sulfurous acid, which would really like to oxidize into the sulfate ion. Once it gets there it's pretty stable in the environment. There's a question of how fast it does that. There are precipitation chemistry measurements that do measure bisulfite in precipitation. It turns out that that can be a pretty important component of the total amount of sulfur that does show up, particularly under conditions that tend to be less acidic. Sulfur dioxide's solubility in water is inversely proportional to the acidity, so you tend to dissolve less SO_2 in water as the acidity goes up. But under conditions where you have moderately acid rain, you have a real good chance of having that sulfur show up as bisulfite, and if you look at the bisulfite contribution to the total sulfur, it's pretty surprising how much there is.

Information Needs – Terrestrial

J. Laurence Kulp
National Acid Precipitation Assessment Program
722 Jackson Place, NW
Washington, DC 20503

The effects on various receptors from the emissions of SO_2, NO_x, and volatile organic compounds (VOC) are determined by the air quality and the acidic deposition at the site. The question I've been asked to address is what information is needed about the source-receptor relationship to explain the terrestrial effects.

Let us begin with the effects themselves. What are the first-order effects on crops and forests that we now understand from ozone, NO_x, SO_2, and acidic deposition at near-ambient levels? "Near-ambient" must be stressed because there has been quite a bit of confusion where effects from experiments with extreme concentrations are not differentiated from those at ambient levels. Also a pollutant may have a negative effect at one type of site and not at another. An example of this is ozone in the stratosphere where higher concentrations would be desirable but at ground level no ozone might be the ideal.

Table I may help in categorizing the terrestrial receptors from a pollutant point of view. There are crops, tree plantations, and natural forests of two types. The effects on crops and on tree plantations are essentially foliar. In these data the soil is managed, pH controlled, nutrients are controlled, and so, in general, soil impacts are not involved.

TABLE I. Terrestrial Receptors

	% of Forest Land	Foliar Damage	Soil
Crops	--	X	--
Tree Plantations	18	X	--
Natural Forests			
low elevation	82	X	X
high elevation	<0.1	X	X

With regard to foliar interaction, the effects are short term, one season for crops and deciduous trees, perhaps as much as four years with conifers. Therefore it is likely that any vegetation will be roughly in equilibrium with its environment, if that environment is not changing rapidly, which indeed is the case with the pollutants under consideration.

Soils may be affected by acidification and nutrient leaching of natural forests over long periods of time unless the buffering action, including mineral weathering is less than the acidic deposition.

Consider first the effects on crops. The dose-response relationship of ozone with a large number of species and

Published 1988 by Elsevier Science Publishing Company, Inc.
Acid Rain: The Relationship between Sources and Receptors
James C. White, Editor

cultivars has been examined over multiple growing seasons. All species of crops that have been tested show a reduction in yield in the presence of ambient levels of ozone if exposed throughout the growing season.

The dose-response relationship is roughly linear in the ambient range but the sensitivities vary widely by species from about 1 percent per year for corn and sorghum to over 20 percent per year for tobacco and lettuce at an average ozone concentration of 100 $\mu g/m^3$ (50 ppbv) at 900-1600 hours during the growing season. As background levels (20-25 ppbv) are approached, some species reach a plateau where little effect is observed on further reduction in ozone concentration.

The effect of increases or reductions in ozone concentrations can be estimated. If ozone concentrations across most of the agricultural land were reduced by approximately 20 percent, a productivity improvement of about 1 percent would result for more resistant species such as corn and sorghum, about 3 to 5 percent for soybeans and about 10 percent in the case of tobacco or lettuce. If all of the crops are weighted by their total value, an average 2- to 3-percent improvement in yield for all agricultural crops is predicted for a 20-percent reduction in ozone. This equates to about a 40-percent reduction in the manmade component of ozone.

SO_2 and NO_x gases also depress the productivity of plants, but higher concentrations than ozone are required. Moreover, in rural areas of the U.S. today, the SO_2 and the NO_x are only a fifth to a tenth the ozone concentrations. Therefore, SO_2 and NO_x may be dismissed from serious consideration as having major impacts on receptors except in certain local situations such as close proximity to a smelter where a thousand parts per billion of SO_2 occurred such as at Sudbury, Ontario or Ducktown, Tennessee.

In the case of acidic precipitation, many experiments have been tried and essentially all show that in the ambient range of pH of acid precipitation, i.e., pH 4.0 to 5.0, there is no measurable effect. It is interesting to note that while a 20-percent reduction in the ambient concentration of ozone results in a readily measurable increase in productivity, no effect from hydrogen ions is observed in plant productivity over a range of over 1000 percent in concentration. In general, acidity must be increased below a pH 3 before effects begin to be observed. This varies somewhat with the particular species and environmental factors, but these levels are well below the average seasonal acidity recorded for growing seasons in the U.S.

There is no negative soil effect on crops from acidic deposition because of the management of soil. In fact, there are some positive effects. The nitrogen that is deposited as nitrate is a significant positive addition to soil so it may reduce the amount of fertilizer required. In the case of acidic soil or soil that could be made more acidic, the farmer has to pay to add lime to bring the pH to where it's appropriate for his particular crop. So the result of acidic precipitation on crops is a little extra cost for additional lime but a significant benefit from nitrogen. Parenthetically, this doesn't mean it is a net societal good to put NO_x in the air because it has already been shown that ozone which is derived from NO_x is a major pollutant and reduces crop yield due to the foliar effect.

Turning to the tree plantations, which represent about 18 percent of the forest land area in the eastern United States, they produce about 50 percent of the wood for lumber and paper. In the long term in the United States, it is expected that by the year 2000 the tree plantations will account for as much as 70 percent of the wood harvested. The natural forests will be used more and more for recreation because, in effect, they are not very efficient for wood production in terms of the utilization of the soil and solar energy. Plantations can be designed to produce the maximum amount of a particular type of fiber or lumber per hectare per year.

In tree plantations, wood is just another crop except that it has a 30-year instead of 1-year rotation. There is no reason to believe that the foliar effect of ozone or acidic deposition is different from agricultural crops. Seedlings of important species are being subjected to dose-response experiments both by individual pollutants as well as mixtures. The data up to the present show the same situation as for crops, i.e., that seedlings are negatively impacted by ozone. They tend to be on the less sensitive side, more like corn and sorghum than tobacco, i.e., yield loss of 1-2 percent per year at ambient levels. Likewise, no significant soil effect is expected, indeed none has been found so far in plantations. If the soils were depleted in nutrients by acidic leaching the soils would be treated with lime to mitigate the effect. The forest industry is a sufficiently sophisticated industry to manage the composition of the soil. They put nitrogen and phosphorus in most plantations so that a complement of lime would require only a small incremental cost. However, most plantations are developed on reasonably good soils which have adequate buffering capacity to neutralize current levels of deposition. Further, virtually all forest land at low elevations are nitrogen deficient. Nitrogen therefore is added in the tree plantations. The amount is commonly ten times the amount that comes down from acid rain in the southeast United States where most of the plantations reside so, for these forests, it is a benefit.

The low elevation natural forest which represents nearly all of the nonplantation forests would be expected to react to ambient ozone and acidic deposition much the same as with crops and with tree plantations. Reported reduction in growth of some of these forests in the southeast and New England in recent decades is probably related to the impact of higher levels of ozone than existed formerly.

Unlike plantations, however, the soils of the natural forests are not managed. Most of them have abundant buffering capacity to acidic deposition but there is a wider range in their chemical composition and a small percentage have low buffering capacity. There are forest soils in some locations that are low in key nutrients such as potassium and magnesium. Probably in some of those locations an interaction with acidic precipitation occurs which is negative. Experiments are now in processs to quantify this effect and establish the area of those soils. It is expected that negative soil effects will be observed on a very small fraction of forested areas and be virtually nonexistent on high productivity land.

Finally, consider the critical, sensitive forested area above the cloud base. This situation is found on the higher mountains of the Appalachian chain above 4000 feet in the northeast and above 5500 feet in western North Carolina. Such forests appear to be in serious decline and show visible damage whether in the Black Forest of Germany, the foothills of the Alps, the higher peaks in the Adirondacks, or in the southern Blue Ridge.

The situation in forest above cloud base is very different from the low elevations. First, these forests are under much more severe natural stresses. The higher the altitude, the greater these natural stresses, and therefore the harder it is for a tree to grow until eventually the timber line is reached where trees can no longer survive. This has nothing to do with pollution. The trees simply cannot cope because of the freezing climate, the high winds, the thin soil, and many other factors.

Second, ozone which retards crop and tree growth at low elevation is present at nearly double the 24-hour average on mountain tops because at low elevations the nitric oxide (NO) generated in the soil and from motor vehicles destroys the ozone at night.

Third, ozone has a greater impact on a plant as the humidity increases reaching a maximum in mists or fog. Therefore in forests above cloud base where mist is commonly present the ozone does more damage than at lower elevations.

Fourth, unique stress above cloud base is the presence of hydrogen peroxide at a concentration of over 1000 parts per billion in the aqueous phase. Hydrogen peroxide is only present in a few parts per billion in the summertime clear air but it is concentrated by solution into the water droplets in the process of cloud formation. It has been shown that hydrogen peroxide also attacks plants, being a vigorous oxidizing agent. In recent experiments in mist chambers, seedlings exposed to controlled hydrogen peroxide concentrations were observed to have physiological damage that was similar to damaged trees in the Black Forest.

A fifth possibility is acid deposition. It has been determined from the Mountain Cloud program that from North Carolina to the Adirondacks the average rain carries a pH of 4.2-4.3 whereas the clouds in the summer average about 3.6. Further there is more total loading of hydrogen ion on the soil due to more total rain and mist-drip than at low elevation. Recent experiments simulating the amount, pH and frequency of mist and rain on red spruce seedlings have shown no significant short-term (seasonal) effect. The long-term soil impact, if any, has not yet been assessed so still may provide another stress to the above cloud base forests.

Finally it is known that at very high levels of nitrate and at near ambient levels of ozone, there may be some reduction in frost hardiness. This is not important at low elevations but, at high altitudes if the tree is sensitized followed by a quick freeze, the tree is stressed and damage may occur. NAPAP has an aggressive research program to study the effects of these various stresses. It is complemented by experimental work proceeding in Europe.

The emission sources for these three pollutants i.e., acid, ozone and hydrogen peroxide are distinct.

The acid comes primarily from SO_2, but must be oxidized. The ozone comes primarily from NO_x. The hydrogen peroxide is derived from volatile organic compounds. Therefore the relative importance of the stress must be defined in order to develop rational economic control strategy.

The source-receptor relationship as normally considered -- emissions, atmospheric chemistry, and transport to a receptor site -- is not sufficient to define the problem. The relative sensitivity of the receptor is also critical. Thus, whether a plant is sensitive to ozone at the rate of 1 percent per year or 10 percent per year at ambient levels makes a difference in the importance of control at a dominant source for that receptor. Another factor is the local meteorology which determines the moisture level, the frequency of fog, the amount of rain, and the type of rain. These parameters are local receptor-related and are independent of long-range transport. Other local factors are elevation and topography. Compare the Los Angeles Basin with Chicago. Both of these areas have the same density of cars, but the problem in Los Angeles is more serious because of a closed topographic rim and the tendency toward air stagnation, whereas in Chicago there is virtually a flat plain with frequent significant winds.

The local meteorology also complicates matters by bringing together air masses with diverse oxidant-to-SO_2 ratios. It may significantly change the ratio of ozone to hydrogen peroxide as well.

Finally the impact of alkaline dust is important as a neutralizing agent for acid rain. Thus the average pH of rain in the Great Plains of the U.S. and the Gobi Desert in China and in other arid parts of the world lies above that (less acidity) of rain in clean pristine areas of the earth.

Turning to the source-receptor influence itself, it is informative to contrast the ozone problem with the acid problem. As noted above, the ozone concentration across the agricultural land in the United States is essentially uniform, i.e., plus or minus about 20 percent of the 7-hour growing season average. This is not because emissions are uniform, but because there is a rather long residence time in the atmosphere for ozone and there is enough mixing over regional distances both horizontally and vertically.

From Minnesota to Virginia, Georgia, New Hampshire, or eastern Colorado, the ozone impact is similar. Therefore, as Ralph Perhac said earlier in the meeting, source-receptor is not that important for ozone in regional effects on crops. Any attempt at control would have to be a very broad scale activity. However, any local reduction targeted at health in urban areas would have some beneficial result regionally.

Peaks of ozone from urban plumes which occur on certain days in the summertime can have a more significant impact on local or mesoscale areas downwind in these areas. Thus for the same total dose over a 7-hour growing period, if a spiked pattern occurs the result is a greater amount of damage on crops than for the same total dose delivered uniformly.

Up to the present time there has not been much continuous monitoring of air quality in rural areas. Most of the EPA network was set up for health effects in urban areas. Only now is the wet deposition monitoring being complemented by dry deposition/air quality measurements in rural areas. This will identify the frequency and intensity of peaks. But these are second-order effects. The dominant theme is the relative uniformity of ozone concentration on a continental basis.

Now whereas ozone is relatively uniform across the country, the annual weighted acid precipitatioin varies by a factor of eight, from the maximum in western Pennsylvania, western New York State and eastern Ohio to west of the Mississippi. Therefore in considerations of control strategy to influence acidic deposition, the source-receptor relationship is of the utmost importance.

Dr. Likens will be talking about the aquatic situation and where there is clear evidence of the acidification of a small percentage of lakes in the Adirondacks. If rain of pH 4.2 water is deposited in a watershed with very low neutralizing capacity, such as a granite surface devoid of soil, the pH of the water in an enclosed lake will be close to 4.2 also. These sensitive sites must be related to specific sources if efficient control is to be designed to reduce effects.

In conclusion, to understand the source-receptor relationship requires accurate atmospheric deposition models. They must be able to predict seasonal averages on specified receptors from identified sources. The oxidant models must be integrated with acid models to unify the picture for long-range transport. Also the frequency and extent of the ozone peaks and the type and extent of the dry deposition are required. Such models which have been validated will permit estimation of the relation of the emissions from given sources to effects at specified receptors.

Information Needs — Aquatic

Gene E. Likens
Institute of Ecosystem Studies, The New York Botanical Garden
Box AB, Millbrook, NY 12545

INTRODUCTION

One of the main points I'd like to make is about complexity. I think there's been much confusion about the inherent complexity of natural ecosystems and about how this complexity affects the interpretation of impacts of acid deposition and other air pollutants on natural ecosystems. Complexity is the norm for natural ecosystems and given the variety of atmospheric inputs, and their variability in time and space, it is not at all surprising that one individual lake or stream may respond differently than another. What is significant is that so many of these individual aquatic ecosystems have become acidified in response to atmospheric deposition during the last few decades.

VARIABILITY OF ATMOSPHERIC INPUTS

The atmosphere is a chemical "soup" and so are lakes and streams. Some gases are in lower concentration in aquatic systems, but there is a complex mixture of all of the kinds of pollutants that come from the atmosphere to the water, some of which produce undesirable ecological effects. Atmospheric deposition of these pollutants can occur as wet (rain, cloud or fog water, and snow) or dry (gases and particles) inputs. All of these can affect aquatic ecosystems.

An important point in evaluating the effects of acid deposition on lakes and streams, as well as on forest ecosystems, is the pattern in which the pollutant materials are transported from sources to these areas. Analyses of the total national emissions of sulfur dioxide during this century show that early in the century most of the SO_2 was emitted by short stacks and fell to the ground locally. More recently, a much greater quantity has been emitted through higher stacks which has allowed those pollutants to reach areas remote from the sources (Figure 1). The effect of changing stack height on emissions of NO_x is quite different (Figure 2). I believe that stack height is an important factor in evaluating the effects of atmospheric deposition on receptors and we should not think only about the total quantity that was emitted. How do pollutants get from where they are emitted to where they are deposited, and how have these patterns changed with time? More needs to be known. Regulatory considerations for controlling emissions of sulfur and nitrogen oxides depend in part on these different relationships between stack heights and sources (Figures 1 and 2).

Acid Rain: The Relationship between Sources and Receptors
James C. White, Editor

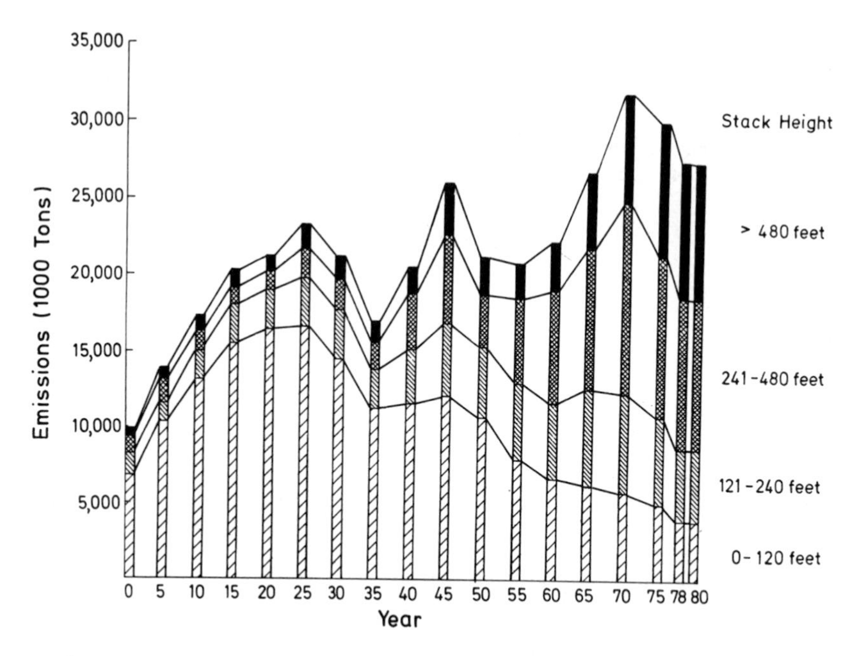

Figure 1. Total U.S. emissions of SO_2 by stack height ranges (source, U.S. Environmental Protection Agency, 1985).

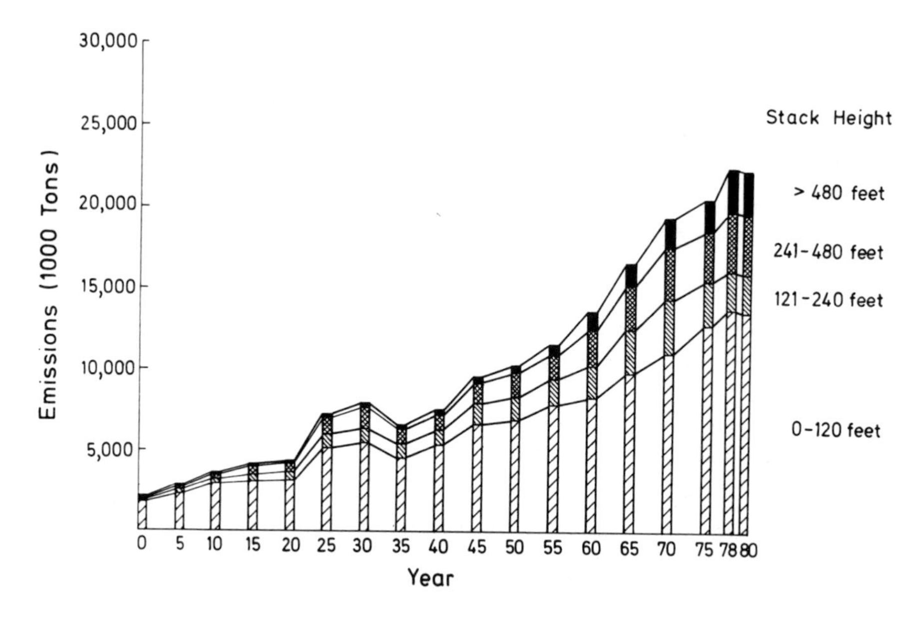

Figure 2. Total U.S. emissions of NO_x by stack height ranges (source, U.S. Environmental Protection Agency, 1985).

Another important factor in source-receptor considerations relates to changes in precipitation chemistry that have occurred over time in North America. We know relatively little about this history (see Likens, 1984a). In an attempt to learn more about background conditions, we have done studies in regions of the Southern Hemisphere remote from industrial or agricultural influences. Then these data were compared with data from eastern North America (Galloway et al., 1984). More recently I compared these data from the Southern Hemisphere with the chemistry of cloud and fog water in the eastern U.S. (Figure 3). There are enormous increases in the concentration of hydrogen ion, sulfate and nitrate in precipitation, and particularly in cloudwater in the eastern U.S. For example, the average pH of cloudwater in the eastern United States is on the order of 3.3 to 3.7. At sea level in Bar Harbor, Maine, we measured values as low as pH 2.4 in cloud water (Weathers et al., 1987).

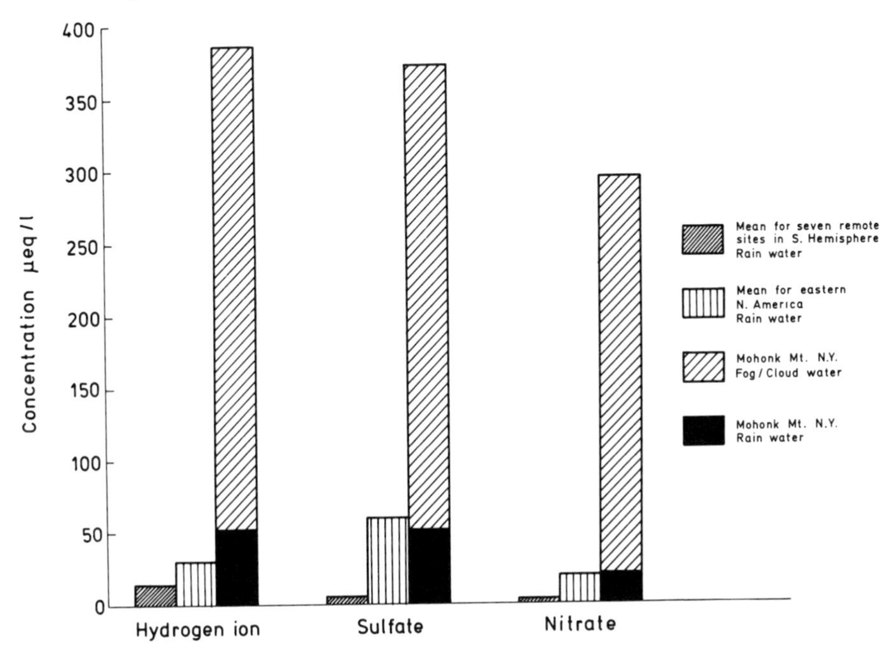

Figure 3. A comparison of the average chemistry of precipitation in remote sites of the Southern Hemisphere, in average precipitation in eastern North America and in cloud and rain water at Mohonk Mountain, New York, during 1984 and 1985 (from Likens, 1987b).

We have calculated what we call "enrichment factors" for precipitation chemistry based on the data from the Southern Hemisphere (Galloway et al., 1984). These enrichments are not uniform throughout the receptor area, but we suggest that there is between eight and sixteen times more sulfur in precipitation in eastern North America than if the atmosphere were not polluted (Figure 4).

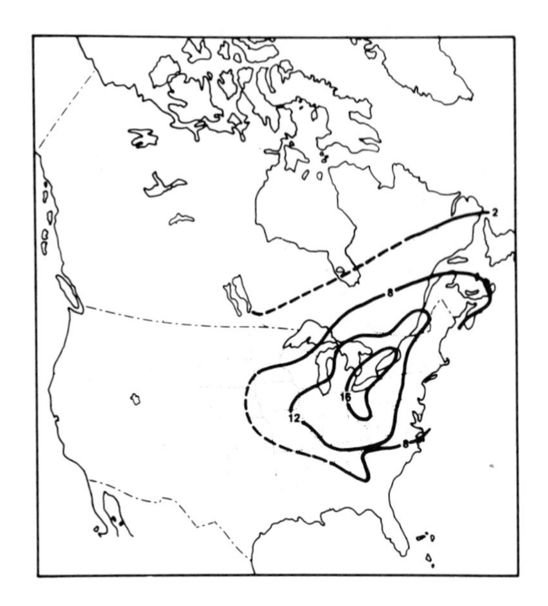

Figure 4. Enrichment of sulfate concentrations in precipitation in eastern North America above that in remote areas of the Southern Hemisphere (from Galloway et al., 1984).

The concentration has changed with time as well (Figure 5). There was appreciable attention given to this change in sulfate concentration in precipitation at Hubbard Brook in the White Mountains of New Hampshire, particularly because this is the longest, continuous record of precipitation chemistry in North America (Calvert, 1983). In terms of the receptors, however, total quantity that is deposited is usually more important ecologically than the concentration.

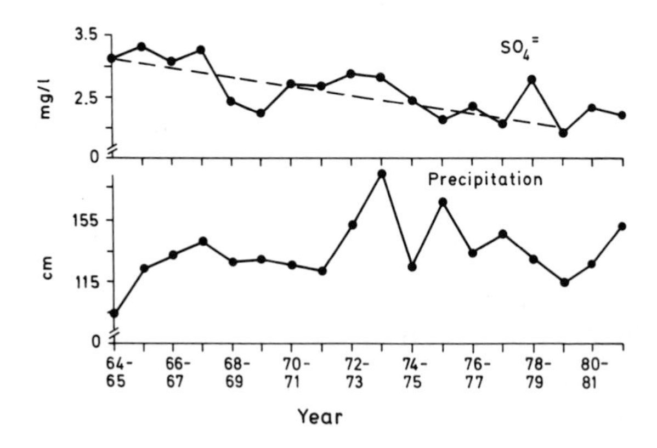

Figure 5. Annual volume-weighted concentration (mg/liter) of sulfate in precipitation and amount of precipitation at the Hubbard Brook Experimental Forest, New Hampshire, from 1964 to 1982. The regression line has a probability for a larger F-value of <0.05.

The amount of sulfate deposited in precipitation at Hubbard Brook has varied annually, but regression analysis shows a statistically significant decline during the past two decades (Figure 6). The most important point, if the threshold were about 10 kilograms sulfate per hectare (Dickson, 1983; Gorham, et al., 1984), is that deposition of sulfur in the northeastern U.S. is about two to four times greater than what is thought to cause chemical and ecological change in sensitive aquatic systems. Annual bulk precipitation input of sulfur has ranged to more than 50 kilograms per hectare during the past two decades at Hubbard Brook. If dry deposition were included (Likens et al., 1977; Eaton et al., 1978; Eaton et al., 1980), the total deposition of sulfur at Hubbard Brook exceeds acceptable levels by three- to sixfold.

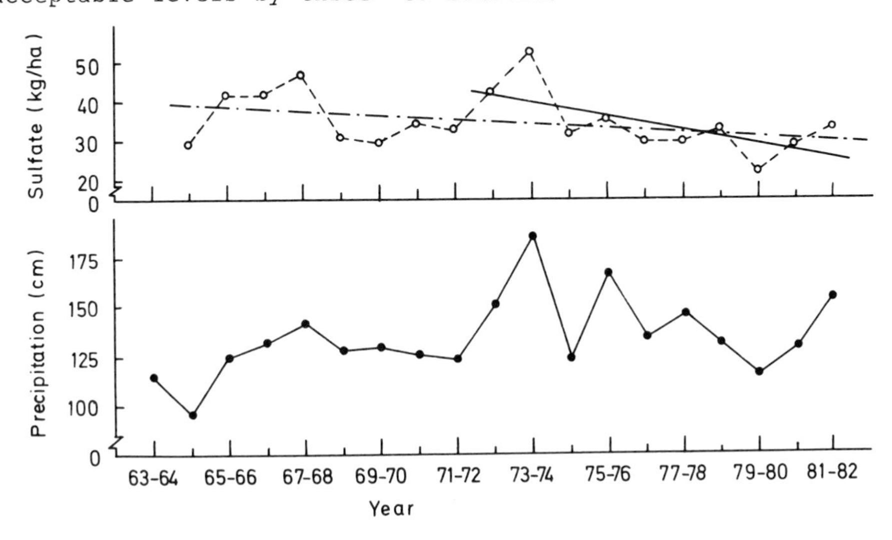

Figure 6. Annual deposition (kg/ha) of sulfate in precipitation and amount of precipitation at the Hubbard Brook Experimental Forest, New Hampshire, from 1964 to 1982. The regression lines have a probability for a larger F-value of <0.05.

There are numerous other components of complexity. Deposition not only changes from year to year, but it changes during the course of a year. For example, sulfate concentration is higher in the summertime and lower in the wintertime at Hubbard Brook, somewhat counter to what might be expected. Streamwater concentrations of sulfate are higher than precipitation and relatively constant throughout the year, mostly because of evapotranspiration which distills off the water, leaving behind the sulfur to drain from the watershed in the stream (Figure 7). I'll come back to this point.

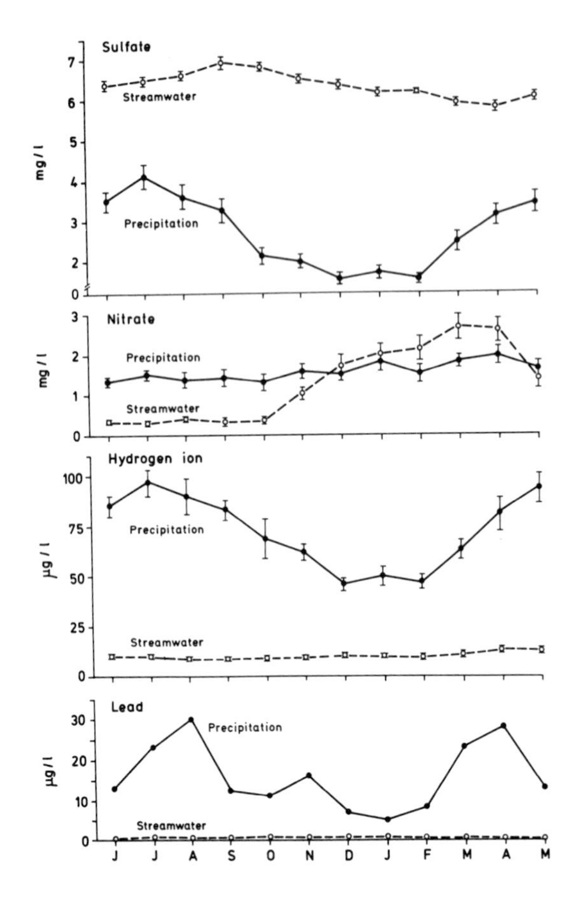

Figure 7. Average monthly concentrations of sulfate, nitrate, hydrogen ion and lead in precipitation (●--●) and streamwater o--o) in the Hubbard Brook Experimental Forest. Vertical bars show one standard error of the mean (from Likens, 1984b).

Nitrate is different ecologically than sulfate, and is very important in natural ecosystems. Because it may be in short supply in forested ecosystems of the northeastern U.S. during the summertime, streamwater concentrations may be very low and precipitation may be an important source for these ecosystems. In the wintertime the opposite occurs; more nitrogen drains from the system in streamwater than enters in precipitation (Figures 7 and 8).

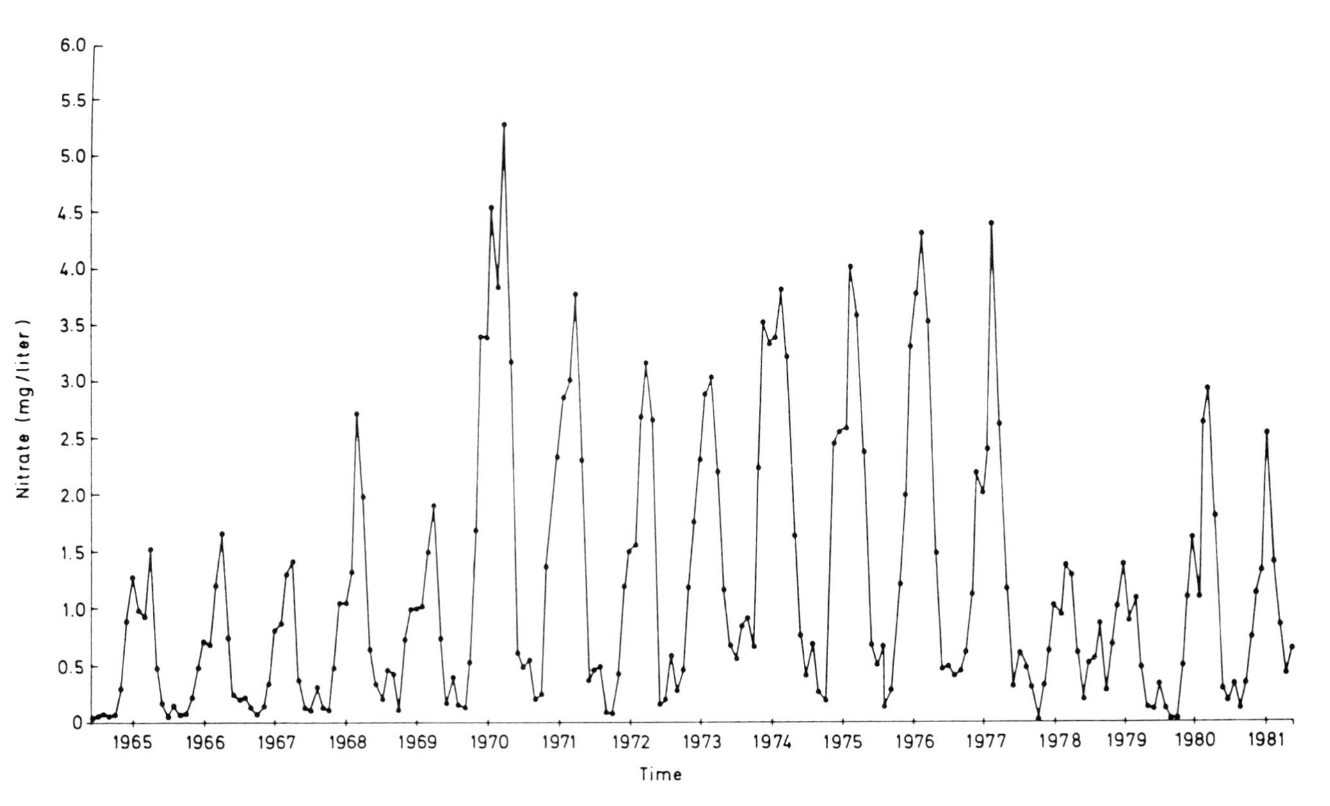

Figure 8. Monthly concentration of nitrate in streamwater from Watershed 6 of the Hubbard Brook Experimental Forest (from Likens et al., 1985, used with the permission of the publisher and the authors).

Hydrogen ion shows a different seasonal pattern (Figure 7). The concentration in precipitation is consistently much higher than that in streamwater, showing neutralization within the ecosystem. Lead has a similar pattern to hydrogen ion. There are much higher concentrations in precipitation than in streamwater. At Hubbard Brook, the terrestrial system is a strong filter for lead and very little actually gets into streams and lakes downstream.

Some rain or snow falls directly into a lake while some falls on the watershed where it may evaporate or drain into the lake. The effective acid neutralizing capacity (ANC) of the entire system is a function of these relative areas receiving acid deposition, and for the lake it is a function of the relative contributions of direct precipitation and drainage to the volume of the lake. Thus, ratio of lake area to watershed area is a major factor in determining the acidification rate of a lake. In some areas, the ANC of the watershed may be large, or at least may control the acidification of low alkalinity lakes. However, Schindler and others (1986) found that the annual, areal generation of ANC in Rawson Lake in the Experimental Lakes Area of Ontario, Canada, was about 4.5 times greater than the areal rate in the watershed for the lake, suggesting major aquatic sources of alkalinity.

The ANC of the terrestrial watershed is determined by the chemical weathering of minerals and ion exchange reactions in the soil. Thus, the flow paths of water draining through the watershed, and particularly the soil profile, adds additional complexity to attemps to predict ANC.

The mobility of the anion accompanying the hydrogen ion in rain or snow is important for determining the acidification of surface waters. The mobility is dependent upon biological assimilation and storage, chemical adsorption and anion exchange in the terrestrial ecosystem. Reuss and Johnson (1985) have suggested a mechanism whereby atmospheric deposition of mobile anions, such as $SO_4^=$, to acid soils can increase the ionic strength and significantly decrease the alkalinity of soil solutions. They proposed that these effects would be instantaneous and reversible if no anion adsorption occurred, but would be delayed by anion adsorption-desorption processes. Thus, sulfate adsorption characteristics in the soil or biotic uptake and storage within a watershed could be critical factors in the acidification of streams and lakes.

We found the output of $SO_4^=$ in streamwater at Hubbard Brook to be directly related to the input of $SO_4^=$ in precipitation (Figure 9). Essentially, all of the sulfur deposited from the atmosphere flowed through the terrestrial ecosystem and was output in drainage water. The close relation between inputs in precipitation and outputs in streamwater would suggest that soils at Hubbard Brook have little capacity currently to adsorb additional amounts of sulfur deposited from the atmosphere. Nodvin and others (1986) have shown that small changes in the pH of the soil solution, however, have a large effect on the sulfate adsorption characteristics of the mineral soil at Hubbard Brook.

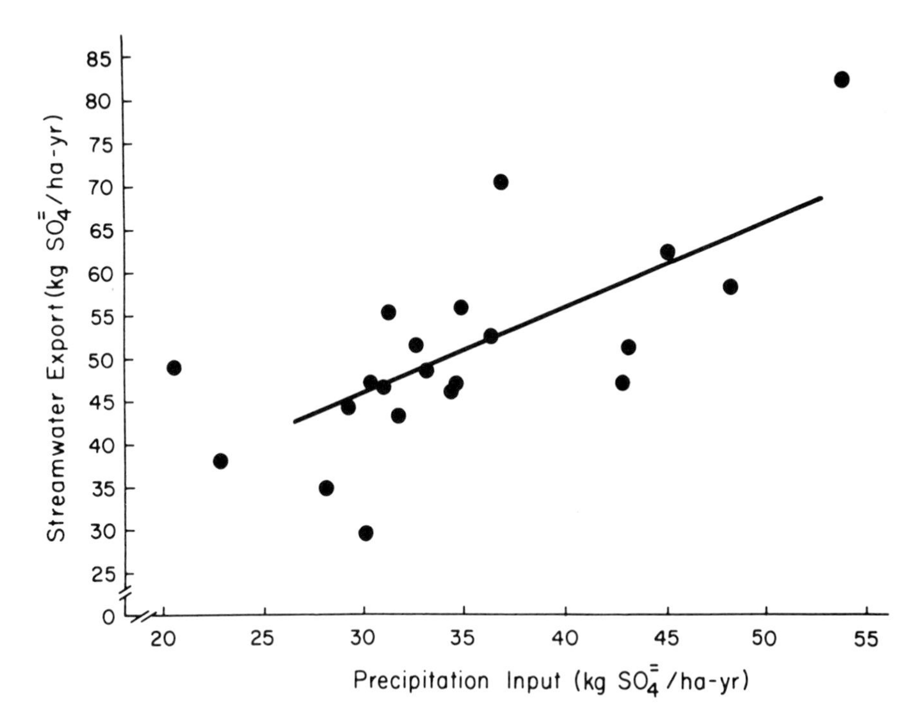

Figure 9. Relationship between the input of sulfate in bulk precipitation and export in streamwater for Watershed 6 of the Hubbard Brook Experimental Forest, New Hampshire, during 1964-65 through 1985-86. The regression line has a probability for a larger F-value of <0.005 and a correlation coefficient=0.69 (from Likens, 1987a).

Galloway (1986) has constructed a model to explain the differences in sulfur balance between the northeastern and southeastern U.S. (Figure 10). The arrows from above represent wet and dry deposition of sulfate. The larger boxes represent terrestrial ecosystems, and the inner boxes in both instances represent the capacity of the soil to adsorb and hold sulfate within the system. In the northeastern United States the adsorption capacity for sulfate in soils is generally filled or nearly filled, so the sulfur input from the atmosphere largely leaves the system in drainage water. This apparently is what is happening at Hubbard Brook (Figure 9).

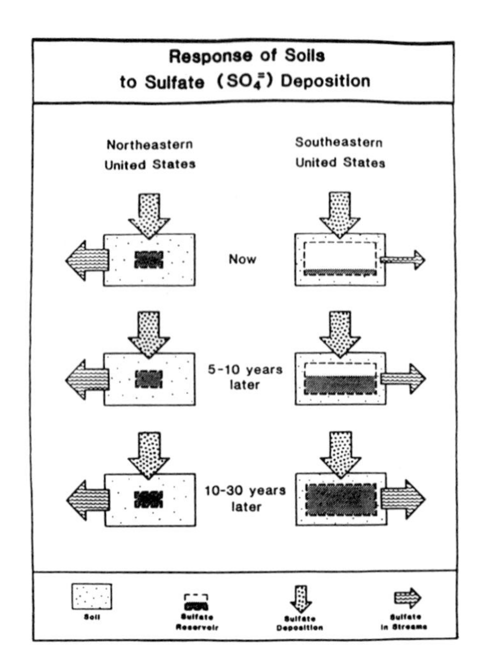

Figure 10. Model for the differential response of soils to sulfate deposition in the northeastern U.S. and in the southeastern U.S. (from Galloway, 1986).

For the southeastern United States, Galloway (1986) has suggested that the sulfur adsorption capacity in soils is much greater. The soils are deep and highly weathered. They are not saturated with sulfur even though they may have had a history of deposition similar to the northeastern U.S. since the beginning of the century. As the adsorption capacity is reduced, the amount of sulfate exported from the system can increase. Recent studies done by Galloway and his colleagues (1987) would indicate that this response is now starting in streams in the southeastern United States. This idea has been called the "delayed response" to acid precipitation. It's been thought by some that this prevents the acidification of surface waters. In fact, it doesn't prevent the acidification, it merely delays the point where acidification can occur at a more rapid rate.

OTHER DISTURBANCES

It has been suggested that if the forested system is disturbed by cutting or fire, long-term changes in acidity will be produced similar to what has been observed in the Adirondack lakes or in other places around the world. We have done a variety of experiments at Hubbard Brook, started in 1965, in which we experimentally cleared or harvested forests. We saw an effect on streamwater chemistry as a result of these disturbances (Figure 11) by comparing the experimentally disturbed forest with an adjacent, forested watershed that was not disturbed. As a result of the disturbance there was an increase in nitrification in the watershed. Both nitrate and hydrogen ions were produced through this process of nitrification. Streamwater became more acid, almost a full pH unit more, but the effect didn't last very long (see Likens, 1984b). In only a few years, the streamwater returned to pre-disturbance conditions (Figure 11).

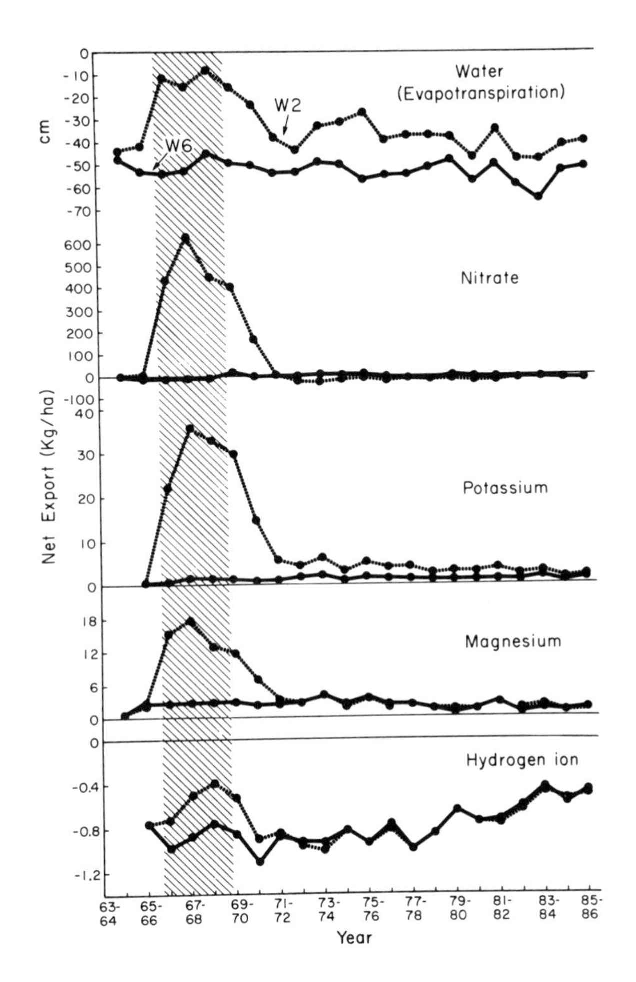

Figure 11. Response of Watershed 2 (●--●) of the Hubbard Brook Experimental Forest to deforestation in comparison with an adjacent, forested Watershed 6 (●—●). Watershed 2 was maintained in a deforested condition for three years (the cross-latched area).

ACID DEPOSITION AND SENSITIVE ECOSYSTEMS

If acid precipitation is added to different kinds of ecosystems, different responses may occur. An example prepared by Dr. Richard Wright from Norway (Figure 12) compares an area receiving acid rain with one that does not. In Birkenes, Norway, precipitation averages about pH 4 and contains high concentrations of sulfate, which fall on coniferous forest. The water from the precipitation flows through very acid soil, pH 4.3 to 5.3, and appears in drainage water, which is quite acid, with high concentrations of sulfate and high concentrations of dissolved aluminum. In contrast, the site in Alaska receives relatively low acid inputs which also fall on coniferous forest with soils even more acid than those in Norway, but the drainage water comes through with much lower concentrations of acid and without high sulfate concentrations or high concentrations of dissolved aluminum.

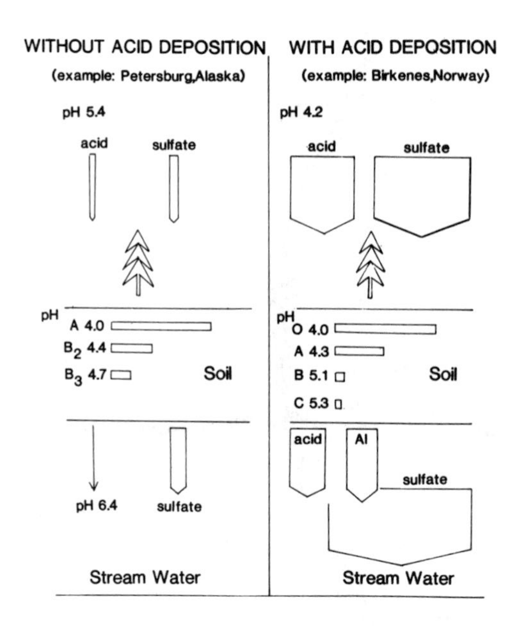

Figure 12. Comparison of rainfall chemistry and streamwater chemistry in areas with and without acid precipitation (from R. Wright, personal communication).

Likewise we (Driscoll et al., 1987) compared data from Hubbard Brook (with acid rain) with the Jamieson Creek watershed in British Columbia (without acid rain). In British Columbia, without the input of acid precipitation, the streamwater had low concentrations of sulfate, very low concentrations of inorganic aluminum and the primary acids are organic acids. Whereas with acid precipitation at Hubbard Brook, the primary acid in streamwater was sulfuric, with very low concentrations of organic acids, and high concentrations of dissolved inorganic aluminum. The presence of inorganic aluminum is particularly significant. In British Columbia the aluminum was in the non-labile, organically-complexed form, and is relatively nontoxic.

At Hubbard Brook the dissolved aluminum is all in the labile, inorganic form, which is highly toxic to organisms (e.g. Baker and Schofield, 1980). These are the kinds of complex responses by natural ecosystems to the loading of atmospheric acids and sulfate. These responses were not due to differences in land use at these sites.

INDIVIDUALITY VERSUS REGIONAL RESPONSE

Another point of complexity: lakes have individual characteristics and respond individualistically to external perturbations. For example, two lakes side by side in the Adirondack Mountain Region may differ considerably. One may have the color of coffee, the other one may be crystal clear. All lakes in a region are not the same; some are deep, some are shallow, some are surrounded by coniferous vegetation, some are surrounded by deciduous vegetation. All of these factors are important in terms of response to an external stress such as acid deposition. And the responses are usually complex and variable. A great deal is made of this variability in the news media and by politicians, although it is not really surprising from a scientific point of view.

Several graduate students and I looked carefully at all of the lakes in the Adirondack Mountain Region where we could compare alkalinity data from the 1930s with values from today (Asbury et al., 1987). A large majority of lakes studied lost alkalinity; that is they became more acid. But there was a spectrum of responses. A few of the lakes even became more alkaline than they were before. Analyzing the lakes on a watershed basis within the Adirondacks Region provided more consistent patterns. The variability depends in part on the terrestrial watershed, the depth of the lake and a variety of other factors that add complexity of response (see Likens, 1987a). Nevertheless, of the lakes studied in the Adirondacks, a highly statistically significant number of them became more acidic since the 1930s (Asbury et al., 1987).

Donald Charles and his colleagues have looked at the history of diatoms in lakes in the Adirondack Mountain Region (Charles, 1985; Charles and Norton, 1986), as others have done elsewhere, to infer the history of acidification. Reconstructing the history of acidity of a lake, such as Big Moose Lake in the Adirondacks, can be done on the basis that the diatoms found in the sediment could only have grown under certain environmental conditions. From a profile of such information in the sediments, it is possible to infer what the conditions were at the time the diatoms were living. From about 1800 until about 1860, Big Moose Lake had a pH between 5.5 and 6.0. The first settlers came in, there was logging and burning, but no effect was seen on the lake as judged by the diatom remains. Again, there was more logging and burning around 1920, but there was no change in pH until about 1950-1955 when the lake rather quickly became very acid. Currently its pH is about 4.5 and the lake has lost many viable populations of fish (Charles, 1984). This pattern has been seen in a large number of lakes that are sensitive to acid deposition, especially in areas of hard bedrock geology with thin acid soils, and are receiving acid deposition (Norton, 1986).

In some lakes, particularly in southern Sweden, a layer of algae may cover the sediments much like a carpet. This mat of algae can be picked up and held in your hand. This layer of algae, or Sphagnum moss in other lakes, may reduce the flux of nutrients to the overlying water, and the waters may tend to become clearer and more acid as a result (Grahn et al., 1974). Other changes in the sediments as a result of atmospheric pollutants (e.g. input of metals, acids, sulfur) may alter the rate of acidification in individual lakes in complex ways (see Likens, 1987a).

Also there are episodic changes. For example, many lakes are located in regions were snow is a prime factor in the hydrologic cycle. Snow accumulates during the winter season and then melts rather suddenly. When snow melts, the first part of the snowmelt water can be much more acidic than the last part, possibly a full pH unit lower. Hultberg (1976) studied a Swedish lake and found that the snow accumulated on the surface of the lake in the range of pH 4.0 to 4.4. Then during a January thaw, the meltwater that drained into the lake had a pH of 3.5 or so just below the ice cover; even at a depth of one meter below the ice it was still appreciably more acid than the snow. Such episodic, seasonal inputs of acidity to aquatic ecosystems may be critical to the survival of organisms.

Aquatic ecosystems contain organisms of great species complexity, e.g. algae, zooplankton, benthic invertebrates, macrophytes, fish, bacteria and fungi, all with variable responses to pH ranges and changes. Some are very sensitive, some are not. Episodic inputs of acids and other pollutants affect different organisms at different levels and at different times, all adding to the complexity of the response for individual aquatic ecosystems.

RECOVERY AND REGULATION

If emissions of acidifying pollutants were reduced significantly, then most systems should stop deteriorating and reverse the trend. How long would it take for systems to respond to reductions in emissions? One of the greatest needs in dealing with this source-receptor issue is to understand more about the complexity of natural ecosystems. It seems to me that the main point is not whether aquatic ecosystems will be acidified, but how fast. At what rate will they continue to be acidified? The complexity of natural ecosystems is very important in understanding what can be expected regarding the source-receptor relationship on a regional scale.

At Hubbard Brook we made an effort, using fifteen years of data, to construct a detailed hydrogen-ion inventory for the entire ecosystem. We analyzed all of the inputs, all of the outputs, all of the internal transfers that we could measure. Currently the pH of headwater streams is about 4.9. Based on the inventory of all protons in the system, we can model what effect a decrease or increase would have in the loading of sulfuric acid, nitric acid or their combinations from the atmosphere. The model shows that, if we took away all of the acid precipitation at Hubbard Brook, the headwater streams would have a pH of about 6.5 or so (Figure 13). Changes in nitric acid loading have relatively little effect, while sulfuric acid shows a much larger effect. At least for this system, the input of sulfur is apparently much more important to long-term acidification than is nitric acid.

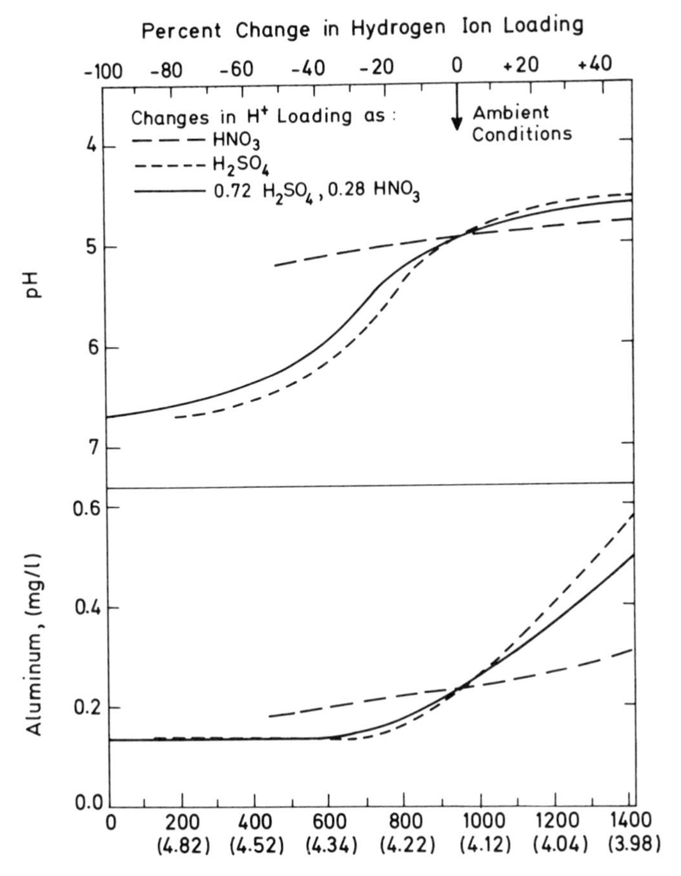

Figure 13. Projected changes in annual volume-weighted mean streamwater pH and aluminum concentration for changes in hydrogen ion loading from precipitation at the Hubbard Brook Experimental Forest, NH. These changes were calculated as if the change in loading occurred as only nitric acid, only sulfuric acid and a combination of nitric and sulfuric acid corresponding to the ambient stoichiometry of Hubbard Brook precipitation. The lower limits of decreases in HNO_3 and H_2SO_4 loading are governed by average concentrations of nitrate and sulfate in ambient precipitation. The abscissa is expressed as hydrogen ion loading in precipitation (eq H^+/ha-yr), ambient precipitation pH and percent change in hydrogen ion loading from precipitation. (Modified from Driscoll and Likens, 1982).

Much is known about the acidification of aquatic ecosystems (see Norton, 1984; Likens, 1987b). There have been thousands of individual papers and several major reviews published on the subject (see Driscoll et al., 1985), but still there is much to learn. To understand how changes in chemistry are related to biological damage we need to know how the rate of chemical change is dependent upon the input of sulfur, nitrogen, hydrogen ion, base cations and metals. These inputs vary seasonally and over the long term.

We need to know about lakes and streams themselves, what their biological and chemical characteristics are, how deep they are, what kind of watersheds they have, what kind of alkalization potential there is within the terrestrial watershed and within the lake itself. We need to know about watershed characteristics. We need to know about base cations in soils and of the ability to mobilize aluminum from those soils. We need to know about sulfur adsorption and desorption in those soils.

We know from what has happened in the Sudbury area, where emissions of sulfur have been reduced recently, that lakes will respond and recover chemically and they do so rather quickly (Dillon et al., 1986). Forests will respond much slower probably and soils, in my opinion, very much slower. Aquatic systems, because of the normal flushing that occurs within them, have the potential to respond quickly, first chemically and then biologically.

The experimental work on acidification of lakes that David Schindler and colleagues have done in Canada shows that biological response may be quick and extensive (e.g. Mills and Schindler, 1986). Natural systems do respond, and this is the principal justification for reducing emissions. When we know more about ecosystems and their natural complexity, and how that natural complexity affects rates of change, we can refine our estimates about what will happen when we cut back the deposition to 10 kilograms of sulfate per hectare per year. But a major reduction is necessary if we want to protect these sensitive natural ecosystems, which are our life support systems.

ACKNOWLEDGMENTS

A contribution to the Program of the Institute of Ecosystem Studies of The New York Botanical Garden and to the Hubbard Brook Ecosystem Study. Financial support was provided by the Mary Flagler Cary Charitable Trust, the Andrew W. Mellon Foundation and the National Science Foundation. The Hubbard Brook Ecosystem Study is a long-term multidisciplinary study initiated in 1963 by F.H. Bormann and G.E. Likens, in cooperation with R.S. Pierce of the USDA Forest Service. Major credit for helping to initiate and sustain the study is also due to J.S. Eaton and N.M. Johnson. The Hubbard Brook Experimental Forest is operated and maintained by the USDA Forest Service, Broomall, PA.

LITERATURE CITED

Asbury, C.E., M.D. Mattson, F.A. Vertucci and G.E. Likens. 1987. Acidification of Adirondack lakes. Submitted for publication.

Baker, J.P. and C.L. Schofield. 1980. Aluminum toxicity to fish as related to acid precipitation and Adirondack surface water quality. pp. 292-293. In: D. Drablos and A. Tollan (eds.). Ecological Impact of Acid Precipitation. SNSF Project, As, Norway.

Calvert, J. (chairman). 1983. Acid Deposition: Atmospheric Processes in Eastern North America. A Review of Current Scientific Understanding. National Academy Press, Washington, DC.

Charles, D.F. 1984. Recent pH history of Big Moose Lake (Adirondack Mountains, New York, USA) inferred from sediment diatom assemblages. Verh. Internat. Verein. Limnol. 22:559-566.

Charles, D.F. 1985. Relationships between surface sediment diatom assemblages and lakewater characteristics in Adirondack lakes. Ecology 66(3):994-1011.

Charles, D.F. and S.A. Norton. 1986. Paleolimnological evidence for trends in atmospheric deposition of acids and metals. pp. 335-506. In: J. Gibson (chairman). Acid Deposition: Long-Term Trends. National Academy Press. Washington, DC.

Dickson, W. 1983. Water acidification -- effects and countermeasures. Summary Document. pp. 267-273. In: Ecological Effects of Acid Deposition. National Swedish Environment Protection Board, Report PM1636, Stockholm.

Dillon, P.J., R.A. Reid and R. Girard. 1986. Changes in the chemistry of lakes near Sudbury, Ontario following reductions of SO_2 emissions. Water, Air and Soil Pollut. 31:59-65.

Driscoll, C.T. and G.E. Likens. 1982. Hydrogen ion budget of an aggrading ecosystem. Tellus 34:283-292.

Driscoll, C.T., J.N. Galloway, J.F. Hornig, G.E. Likens, M. Oppenheimer, K.A. Rahn and D.W. Schindler. 1985. Is there scientific consensus on acid rain? Excerpts from six governmental reports. Ad Hoc Committee on Acid Rain: Science and Policy. October 1985. Special Publication of the Institute of Ecosystem Studies, The New York Botanical Garden, Millbrook, New York. 13pp.

Driscoll, C.T., N.M. Johnson, G.E. Likens and M.C. Feller. 1987. The effects of acid rain on the chemistry of headwater streams: a comparison of Hubbard Brook, New Hampshire with Jamieson Creek, British Columbia. (Submitted)

Eaton, J.S., G.E. Likens and F.H. Bormann. 1978. The input of gaseous and particulate sulfur to a forest ecosystem. Tellus 30:546-551.

Eaton, J.S., G.E. Likens and F.H. Bormann. 1980. Wet and dry deposition of sulfur at Hubbard Brook. pp. 69-75. In: T.C. Hutchinson and M. Havas (eds.) Effect of Acid Precipitation on Terrestrial Ecosystems. Nato Conference Series 1:Ecology 4. Plenum Publishing Corp.

Galloway, J.N. 1986. Testimony to the Committee on Environment and Public Works, U.S. Senate, Washington, DC, October 2, 1986.

Galloway, J.N., G.E. Likens and M.E. Hawley. 1984. Acid precipitation: natural versus anthropogenic components. Science 226:829-831.

Galloway, J.N., B.J. Cosby, G.M. Hornberger, P.F. Ryan and E.B. Rastetter. 1987. Delayed reactions of two southeastern streams to increased atmospheric deposition of sulfur. (Submitted)

Gorham, E., F.B. Martin and J.T. Litzau. 1984. Acid rain ionic correlations in the eastern U.S.A. 1980-1981. Science 225:407-409.

Grahn, O., H. Hultberg and L. Landner. 1974. Oligotrophication -- a self-accelerating process in lakes subjected to excessive supply of acid substances. Ambio 3:93-94.

Hultberg, H. 1976. Thermally stratified acid water in late winter -- a key factor inducing self-accelerating processes which increase acidification. pp. 503-517. In: L.S. Dochinger and T.A. Seliga (eds.) Proceedings of the First International Symposium on Acid Precipitation and the Forest Ecosystem. USDA Forest Service, General Tech. Report NE-23.

Likens, G.E. 1984a. Acid rain: the smokestack is the "smoking gun." Garden 8(4):12-18.

Likens, G.E. 1984b. Beyond the shoreline: a watershed-ecosystem approach. Verh. Internat. Verein. Limnol. 22:1-22.

Likens, G.E. (ed.). 1985. An Ecosystem Approach to Aquatic Ecology: Mirror Lake and its Environment. Springer-Verlag New York Inc. 516 pp.

Likens, G.E. 1987a. Acid rain and its effects on sediments in lakes and streams. In: Proceedings of the Fourth International Symposium on the Interaction Between Sediments and Water. Melbourne, Australia. (In Press)

Likens, G.E. 1987b. Chemical wastes in our atmosphere -- an ecological crisis. In: Proceedings of the International Conference on Industrial Crisis Management, New York University, September 1986. (In Press)

Likens, G.E., F.H. Bormann, R.S. Pierce, J.S. Eaton and N.M. Johnson. 1977. Biogeochemistry of a Forested Ecosystem. Springer-Verlag New York Inc. 146 pp.

Likens, G.E., F.H. Bormann, R.S. Pierce and J.S. Eaton. 1985. The Hubbard Brook Valley. pp. 9-39. In: G.E. Likens (ed.) An Ecosystem Approach to Aquatic Ecology: Mirror Lake and its Environment. Springer-Verlag New York Inc.

Mills, K.H. and D.W. Schindler. 1986. Biological indicators of lake acidification. Water, Air, and Soil Pollut. 30:779-789.

Norton, S. (chairman). 1984. Acid Deposition: Processes of Lake Acidification. Environmental Studies Board, National Academy Press, Washington, DC.

Nodvin, S.C., C.T. Driscoll and G.E. Likens. 1986. The effect of pH on sulfate adsorption by a forest soil. Soil Science 142(2):69-75.

Reuss, J.O. and D.W. Johnson. 1985. Effect of soil processes on the acidification of water by acid deposition. J. Environ. Qual. 14(1):26-31.

Schindler, D.W., M.A. Turner, M.P. Stainton and G.A. Linsey. 1986. Natural sources of acid neutralizing capacity in low alkalinity lakes of the Precambrian Shield. Science 232:844-847.

Weathers, K.C., G.E. Likens, F.H. Bormann, J.S. Eaton, K.D. Kimball, J.N. Galloway, T.G. Siccama and D. Smiley. 1987. Chemical concentrations in cloud water from four sites in the eastern United States. In: Proceedings of the NATO Advanced Workshop on the Acidic Deposition to High Elevation Sites. Edinborough, Scotland. D. Reidel Publishing Co. (In Press)

U.S. Environmental Protection Agency. 1985.

Questions and Answers following Information Needs Session

Thursday, December 4, 1986

Q:

In a previous session, some barbs were thrown towards NAPAP concerning the disarray of the program, lack of reports, etc. Do you have anything to say on that?

A: (Kulp)

I don't want to take the rest of the morning. It's always difficult to answer yellow sheet journalism and heresay. The fact is that the NAPAP program is probably more tightly focused now than it has been in any time in its history. It has grown, of course, and become more complex. This is the year that we define for sure what we are going to accomplish by 1990 in terms of first-order answers. The reports that are referred to are essentially in synch with what they have been in prior years. They are being accelerated. In prior years the annual reports came out about 11 months after the end of the calendar year. This year it will be the twelfth month, but the 1986 report will be out in April (1987), so that we can go on with many kinds of things.

Q:

Has there been, to your knowledge, any corroboration of the findings concerning hydrogen peroxide effects on seedlings?

A: (Kulp)

Not to my knowledge. But in our program we are starting experiments in 1987.

Q:

Can you tell us where they will be done?

A: (Kulp)

At Oak Ridge, for one place. We hope to have at least three places.

Q:

What about this magnesium deficiency theory that has surfaced recently. Any comments?

A: (Kulp)

Well, I think there are certainly some soils that are magnesium deficient, and it is clear that some trees that are stressed in certain areas will manifest this, in yellowing needles and so on. To what extent the acidic deposition is responsible, I think remains to be fully worked out. In Germany, where they have attempted fertilization on these particular magnesium deficient soils, you do get good response. But they tend to be in the transition zone of the damage. When you get to much higher altitude, where there is very severe damage on all species essentially, this treatment doesn't straighten it out. This is part of the picture for certain specific soil types.

Q:

In both of your informative discussions, you haven't really given the modelers in the audience much guidance as to what periods we should be modeling. Are we interested in deposition over a whole year, or in deposition in snow melt, or in peak depositions during the summer, or in ozone, or all of the above? Our source-receptor relationships can be wildly different depending on which period we are interested in.

A: (Likens)

All of the above, obviously. The total deposition to a natural system is one key component. But the frequent episodes, either in snow or in individual rain events in the range of pH 2.1 or anywhere less than pH 3, we need to know about. We need to know when they are going to happen, how to predict them, why they happen. As Mr. Kulp was suggesting, if we get pHs of less than 3, even less than 3.5, you start to see damage.

A: (Kulp)

I think I'd take a slightly different tack on that. As far as foliar growth is concerned, with the ozone relationship, we are interested in the early summer season. I don't think that the rest of the year is that critical. I suspect that it is not realistic to have long-range projections of spikes, whether they are of acid rain or ozone. On the other hand, we do need to understand the fingerprints of various areas that might be reproduced. Let's say, in the case of ozone, consider the peaks, consider the type, because we have seen rural areas where there are very few peaks.

Q:

We're concerned in the west not only with sulfate but also with nitrogen, especially in our sensitive high elevation areas. Has anyone looked at the potential for high elevation lake eutrophication? Would you comment?

A: (Likens)

There are a couple of studies that I know about that are beginning to look at that, but normally the limiting factor for most lakes is the availability of phosphorus. Additions of nitrogen to dilute mountain lakes often can be very important to productivity. And if the lake lacks phosphorus, you can add all the nitrogen you want and you won't produce any eutrophication.

Q:

You implied that, because ozone is so uniformly distributed over most of the U.S., the source-receptor relationship for ozone, whatever it is, is not of much consequence. I'm not sure there is such a thing as a source-receptor relationship for ozone, since ozone is not emitted from a source, but is a secondary pollutant. But if you look at the precursors of ozone, NO_x and VOCs, there is certainly some source-receptor relationship for those. Are we to conclude from your remarks that, if you were to reduce emissions of NO_x and VOCs, let's say preferentially in some areas, that it would not create an ozone hole in the region where you reduce those emissions considerably?

A: (Kulp)

I guess I would say that you wouldn't see a very well defined hole. What we now have is this rather uniform distribution over very large areas, and these areas differ dramatically in emissions. In New York and along the East Coast we have roughly ten-fold the emissions over what you have in Nebraska, and yet the ozone is about the same. I was not indicating that it was not important, nor that there isn't a very broad scale source-receptor relationship. I just doubt that it can be usefully focused in the way that the acid deposition relationship might.

Comment:

I don't want to belabor the point, but I think that you are misinterpreting the information. I want to make three points. First of all, the distribution is broad for sulfur. While the emissions differ very radically over the eastern United States, with sulfate, a secondary pollutant, there is only a difference of a factor of two over a very broad range of cloud forms in terms of sulfate deposition or concentrations. Secondly, a lot of the relative stability or uniformity of ozone is due to a very high natural background. And what you really want to be interested in probably, from a biological point of view, is a change in the incremental ozone due to anthropogenic influence. That number will change much more radically over the problem areas of the United States than will the total ozone. So I think there is less of a difference as far as the acid problem, if you look at sulfate, and more of a difference in ozone if you present it from an anthropogenic standpoint.

A: (Kulp)

Thank you, but I read the data a little differently. I think there is more than a factor of two. Also from what Dr. Likens' slides and others show, it is different from the acidity situation.

Q:

Dr. Likens, you mentioned twice in your talk that, to protect aquatic systems, the loading needs to be somewhere near 10 kilograms per hectare per year. Am I to understand correctly that you are suggesting a halving of the current targeted loading?

A: (Likens)

Yes, I am.

Q:

I'd like to comment on your comment, Dr. Kulp, about the

leaching of nutrients from the soil, and also the variability of the soils. In certain areas we have outwash soils that are left over from the glacial age which are extremely low in potassium. And though it is true that there is not much obvious death of trees at this point, there is a tremendous amount of the yellowing that is a characteristic of potassium deficiency going on in the Adirondack area and Quebec and through that area. I think we want to watch that because it has a cumulative effect. In other works, what is being leached today leaves less for next year and so on. If you look at the soil analysis, there are signs of extraordinarily low levels of potassium in the soil, and it is very soluble, and presumably when this glacial age wash came through it was moving a lot faster, so I think it is something that deserves careful attention.

A: (Kulp)

Thank you. I agree with that. Certainly there are special cases where apparently there's a question.

Q:

I have a question about the chart that Dr. Kulp showed concerning the ambient concentrations of ozone, seven-hour average. You mentioned the problem about ozone monitors now primarily in urban areas; you showed a chart which showed the seven-hour ambient ozone concentrations during the growing season. It's not clear to me where those stations are located. Obviously, there is one at the top of the Adirondacks; that's a rural location. Where were most of the locations when this data was taken?

A: (Kulp)

That was early 1980s work it was based on the existing rural or semi-rural stations. We don't have very much better data now, but these were generalized numbers from a limited set of monitoring stations.

Q:

Dr. Kulp and Dr. Likens have suggested that we need a great deal more research. The NAPAP program will be continuing on through 1990 before we can actually recommend at the first level some sort of accurate limitation of deposition. Dr. Likens, on the other hand, also recommended that we need to know a great deal more about aquatic and terrestrial interaction, both chemical and biological. Yet you recommend a specific target loading rate. Do we need to know more about the complexity of these systems, and quantify these effects before we can recommend such a target loading rate?

A: (Likens)

My response to that is that I may have confused what I was trying to say. It's clear to me that aquatic systems have been acidified. The complexity that I was talking about is the kind of complexity that natural ecosystems inherently have. As a scientist, I want to know more about that, and I've spent most of my academic career trying to find out more about that. In direct answer to your question, I don't think the additional information about that complexity is necessary to make a regulatory action decision. I think that is about as clearly as I can state it. We know a lot already, and there is a lot we don't know. In my opinion, what we don't know is not a requirement for taking regulatory action. Obviously, that is a very different opinion from that of the EPA or the current administration.

A: (Kulp)

I just want to make it very clear that I was not implying in any way that anybody has to wait until 1990 to make a decision. NAPAP has gotten more and more data every year, and every year there is a more rational basis for making a regulatory decision. NAPAP is not involved in recommending anything one way or the other; we are strictly trying to get at the science.

A: (Likens)

Since you've given me a soapbox, may I have 30 seconds more? It is my view as a scientist that the scientific input to decisionmaking such as this is not very important. I would say that only 15 to 20 percent of a decision is based on our scientific understanding, although scientists have a hard time saying that. That's the way I view it. Most decisions are made on the basis of politics and economics. In that regard, about a year ago, some of us put together a little pamphlet called, "Is There Consensus on Acid Rain?" About half the group were policymakers and half were scientists. We came out of it by saying, "Let's look at all the governmental reports and see what they said. Don't worry about what we said; let's see what they said." We did that, just copied out the answers to questions that we generated. The answers came from these governmental reports. If anyone would like a copy, I've brought a few.

Comment:

I'd like to make a comment. I think this conference may go down as the conference where the tropospheric ozone hole was invented. I think the point was well taken that ozone is not a primary pollutant; it is manufactured in the atmosphere. One other thing that I think we need to remember is that it is a manufacturing process that is not linear, and that controlling either hydrocarbons or NO_x may have varying effects depending on their relative abundance and relative ratio.

Q:

I'd like to ask Dr. Kulp and Dr. Likens: given the state of affairs now, and what we know about source-receptor relationships, do you consider that the information we have at hand on modeling could give a fairly concise opinion as to what sources are impacting the east? Are these sufficiently reliable, in your opinion, to warrant and support the kind of proposed legislation that is now on the table in Congress?

A: (Kulp)

I'll pass on that one; it's a policy question.

A: (Likens)

I'm not a policy analyst; as I said; I'm a scientist. I can give you my opinion as a person who walks the street, that if the questions come down to me as, "Do we want to take action or don't we want to take action? What is the cost? Do we want to spend the money or, if we don't want to spend the money, what will be the repercussions?" I'm not in any position to make any of those decisions. If I were, I would take action.

Q:

I recall a remark of Dr. Kulp's that we don't know what is important, but we sure will...

A: (Kulp)

That was regarding high altitude forest.

Q:

Not acid rain research?

A: (Kulp)

Absolutely not. It had only to do with high altitude forest.

Q:

Then that's another question. Do you think it is fair to say that there is little or no effect on the forests when we have not done systematic surveys to find out what has been affected and what has not?

A: (Kulp)

No, and I didn't say that, that there is little or no effect. What I said was that we have good experimental evidence of the ozone effect on a foliar basis. We have a fair amount of experimental evidence that, in most areas, at low elevation there is no major known effect for ambient acidity. I also said that the ambient acidity clearly would affect certain soils. We have a lot of research to do to understand that. In my comments I mentioned one clear case, and when we get to high altitudes, you have a much more complicated situation. So yes, a lot of research remains to be done.

Since Dr. Likens was able to take a little time, maybe I will make a statement before we close our session. I think it is of great importance to a group like this. Technology has emerged in the last 18 months which will totally solve this problem over time, as far as SO_2 emissions are concerned from utility boilers. I feel that we have an accurately documented technology now that will become the technology of the future, and it will do so regardless of what the government does, or regardless of whether there is any further regulation. It will remove better that 99 percent of the SO_2, and better than 97 percent of the NO_x at coal-burning facilities at a cost of less than the current NSPS pulverized coal scrubber technology. Also it won't have all the mountains of gypsum sludge that nobody likes to see coming out. Without changing anything major in the long-term, 50 years, the problem is solved. If it turns out that lakes are dying and soils acidifying - that's one scenario; if we find that it takes many years for any change beyond what we're having now - that's another scenario, and decisionmakers must address the reality.

Let us use the science we have now and use it well; this is the rational approach. The decisionmakers must decide, "Are the quantitatively measured effects and damage such that a more rapid replacement of generating capacity is required?" Normal replacement of generating capacity by technology that essentially has no emissions, will not cost society one net additional dollar. With any retrofit, crash program or not, it will cost whatever it will cost. This has to be measured by decisionmakers against the value of what they are going to get. But in the long term, this problem is solved.

Editorial Note:
Dr. Kulp is referring to the new technology in integrated coal gasification combined cycle combustion.

Simulating Source–Receptor Relationships for Atmospheric Contaminants

Perry J. Samson
Department of Atmospheric and Oceanic Science
Space Physics Research Laboratory
The University of Michigan, Ann Arbor
MI 48109-2143

ABSTRACT

Source-receptor relationships for atmospheric contaminants are estimated frequently through the use of computer simulation models. Unfortunately, the accuracy of these models in estimating source-receptor relationships has never been tested directly because the "true" flow of air over distances greater than a few hundred kilometers is difficult to prescribe. Instead their skill has been inferred from their ability to reproduce temporal and/or spatial patterns of observed ambient concentrations and/or pollutant deposition. Given our inability to verify simulated source-receptor relationships, it behooves us to identify the magnitude of the uncertainties in these simulations and to estimate their possible impact on source-specific control strategies.

This paper presents:

- representative estimates of source-receptor relationships from current atmospheric transport models,
- an analysis of the uncertainty in source-receptor relationships based on uncertainties in model parameters,
- an analysis of the variability in source-receptor relationships based on variability in meteorological conditions, and
- a description of how the next generation of atmospheric transport models will be used to estimate source-receptor relationships.

INTRODUCTION

Legislators and administrators charged with the responsibility of designing air pollution emission control plans to protect our environment have often sought guidance from computer simulation of pollutant movement in the atmosphere. Invariably, such simulations of specific contributions to observed concentration levels have nonetheless raised questions concerning the degree of confidence that should be attributed to such estimates. The uncertainties have arisen because of simplifying assumptions about pollutant dispersion, inadequate measurements of wind speed and direction at the stack, and inhomogeneities in the local terrain, even when relatively simple source-receptor geometries were being addressed. With the recent acknowledgement of the potential role of long-range transport of pollutants on the air and precipitation quality, the

Acid Rain: The Relationship between Sources and Receptors
James C. White, Editor

role of computer simulation of source-receptor relationships has increased dramatically. With the needed increase in sophistication of computer algorithms to describe the meteorological, chemical, and physical processes affecting atmospheric concentrations over whole regions there has been an increasing debate on the reliability of source-receptor estimates generated by models containing a myriad of *possible* uncertainties.

The computer simulation of the transport, transformation and deposition of sulfur species in the atmosphere has been used to infer culpability for observed air quality and deposition patterns on a regional basis, and hence to optimize the costs of emissions control. Unfortunately, while much is known about the chemical and physical processes associated with atmospheric transport and deposition, there are still many uncertainties which can influence the results of these simulations. This paper illustrates the uses and limitations of simulated source-receptor relationships when applied to atmospheric sulfur species by: (1) presenting representative estimates of source-receptor relationships from a simple atmospheric transport model, (2) demonstrating the uncertainty in source-receptor relationships based on uncertainties in model parameters, (3) estimating the variability in source-receptor relationships based on variability in meteorological conditions, and (4) describing how the next generation of atmospheric transport models will be used to estimate source-receptor relationships.

A source-receptor relationship matrix simply describes the linear contribution of given source region to a given receptor region. Mathematically, the matrix can be written as

$$V_i = \Sigma \, a_{ij} \, Q_j \qquad (1)$$

which is is composed of Q_j, the emissions in region j on the right side, and your measurements, V_i, in region i on the left side. The source-receptor relationship[1] matrix a_{ij} represents the *assumed* linear proportionality between what is emitted in region j and what is observed at receptor i. While there are many reasons why the relationship between sources and receptors may not necessarily be a linear relationship, most of the estimates which have been compiled, and most of the legislative action which has been proposed have assumed that there is a linear transfer.

MODEL TYPES

There are a number of techniques which have been used to estimate the source-receptor relationship matrix a_{ij}. Those which rely on computer simulation techniques can be categorized into two basic frameworks. The first are *Lagrangian*-type models, which are available either as a source-orientated model, as illustrated in Figure 1a or as a receptor-oriented model, as illustrated in Figure 1b. The Lagrangian-source model type follows emissions from a source or group of sources downwind, and provides estimates of the contributions of those sources at various points along the path of the air. The advantages and disadvantages of this type of model both lie in its relative simplicity. Models of this type do not require large computer systems and interpretation of their results is generally straightforward. Unfortunately, this model type allows the incorporation of only the most highly parameterized descriptions of the chemical and

---

[1]Also called the unit transfer matrix since it is a normalized description of the rate of transfer from the source region to the receptor region.

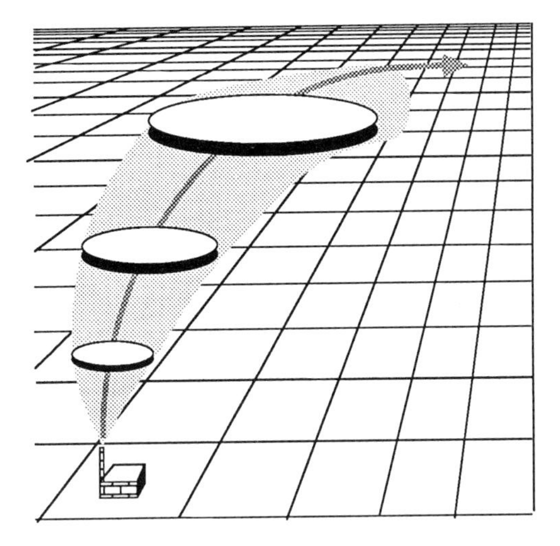

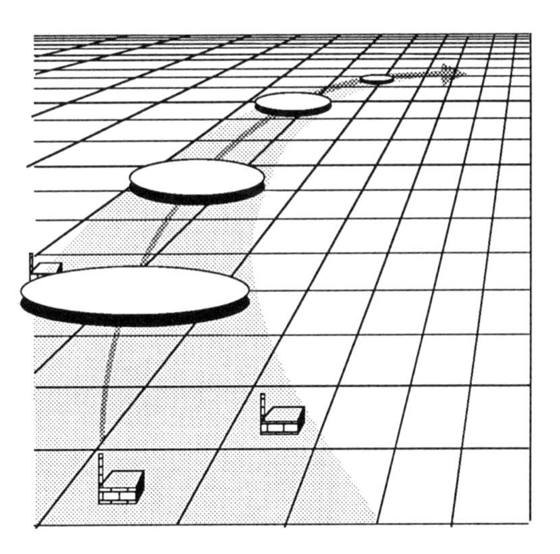

Figure 1. A schematic representation of (a) source-oriented and (b) receptor-oriented Lagrangian air quality models. In the source-oriented model the air is followed downwind of a source or source region dispersing and depositing as it moves. In the receptor-oriented model the probability of upwind source influence on a particular receptor is estimated.

physical processes of the atmosphere. These models also require that we make simplifying assumptions concerning the vertical distribution of pollutants and the rate of pollutant deposition.

Many of the same limitations cited above for the source-oriented Lagrangian model also exist for the receptor-oriented models. With this model type, a trajectory is followed upwind from a receptor to estimate which sources could have contributed to that receptor.

Common to all Lagrangian models is the need to estimate the trajectory of the air. Our knowledge of the movement of the winds is limited because the information we must use for winds above the surface is only available every twelve hours from stations which are separated by about 400 kms. Moreover, we do not know the vertical distribution of the contaminants so we are forced to make assumptions about the relative importance of winds at various height above the ground.

If the contaminant concentration is fairly well mixed through some layer of the atmosphere, a layer-averaged wind will probably represent the mean movement of the mix. But if the concentration is emitted near the surface and is not mixed vertically, it will be moved by the surface winds. Since the distribution in this case is not uniform, a layer-averaged wind will probably not describe the movement of the contaminant accurately.

One way to address this problem is to admit our uncertainty, and to describe trajectories not as a single line but, rather, as probability fields, where the axis of the trajectory represents the axis of highest probability of air being transported to a receptor. The gradient in probabilities to the left and right of the axis of the trajectory will be dictated by the magnitude of the wind shear through the layer. Add to the probability field due to the effects of transport and dispersion alone information about where it precipitated upwind of the receptor and a more realistic probability of influence is obtained. If precipitation occurred upwind of the receptor the probability of air pollutants moving from the source area to the receptor would be decreased upwind of the precipitation. Likewise we can decrease the probability of impact of sources far upwind due to dry deposition. This loss may be at least partially offset by the increase in probability due to chemical production of the pollutant en route.

The types of results which can be obtained are shown in Figure 2. This graph shows the resulting probability field for contribution to sulfur wet deposition in 1978 at Whiteface Mountain. Since this field does not include the actual location of pollutant sources it is referred to as the *natural potential* to pollute. This natural potential is based on the flow of the wind in 1978 to the site at Whiteface and on the precipitation patterns across the United States and Canada during that year. You can see that there was a bias, as the axis of highest possible contribution extended toward the southwest from Whiteface. Hence the sources in the State of Ohio were more likely to contribute to Whiteface sulfur wet deposition than were sources to the northwest, because the meteorology dictated the difference.

When this potential for contribution is multiplied by the exact location of sulfur dioxide emissions you get an estimate of a two-dimensional source-receptor relationship matrix. Based on this analysis [1] the combination of large source strength in eastern Ohio and western Pennsylvania and the southwest bias of natural potential lead to the estimate that the highest portion of sulfur deposition in precipitation at Whiteface Mountain was due to emissions in the State of Ohio.

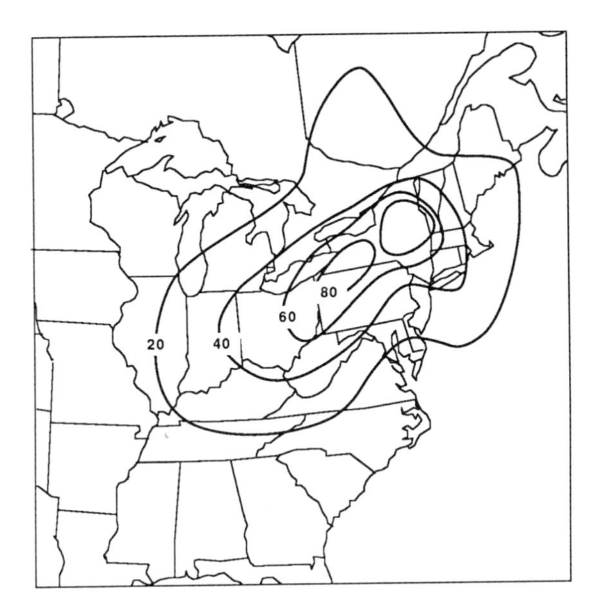

Figure 2. The probability of contribution to sulfur wet deposition at Whiteface Mountain, New York based on the meteorological conditions in 1978. This probability field ignores the actual location of sources and illustrates the potential of upwind regions to influence concentration of sulfur in precipitation.

The evidence that this estimate is correct is quite circumstantial. Because direct measurements of source-receptor relationships are limited, the reliability of estimates of source-receptor relationships are based on the ability to simulate concentration variations at a receptor. We hope that if the model simulates observed concentrations properly that the components of the model are correct. That is a giant leap of faith. It is quite possible that the right answers can be obtained *for the wrong reasons*.

For example, a comparison of simulated daily ambient sulfate concentrations at Whiteface Mountain, New York versus observed concentrations produced a correlation of approximately 0.7. However, the scatter diagram shown in Figure 3 shows that a great deal of variation still remains. The solid lines represent agreement within a factor of two. The correlation simply reflects the model's skill for differentiating days of generally low concentrations versus those of higher concentration. This skill probably only reflects the differentiation of transport from regions of smaller emission sources versus transport from regions of larger emissions. Segregating those days of transport from the Midwestern United States, for example, the variations in observed concentrations are not reproduced by this particular model.

It is scatter of the magnitude shown in Figure 3 that has been used to justify the EPA's reluctance to certify regional-scale air quality models for assessment studies. It should be noted, however, that plots of predicted versus observed concentrations for EPA-approved short-range (<50 km) models, used in promulgating legislation, generally show a similar degree of scatter.

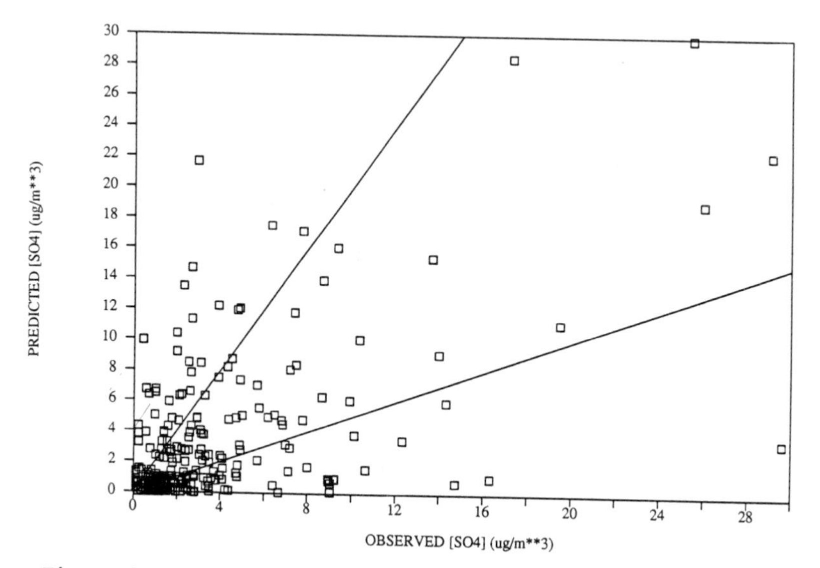

Figure 3. A scatter diagram of predicted versus observed daily average concentrations of sulfate at Whiteface Mountain, New York using a Lagrangian receptor-oriented model developed by Samson and Small [1]. The solid lines indicate factor of two differences between predicted and observed.

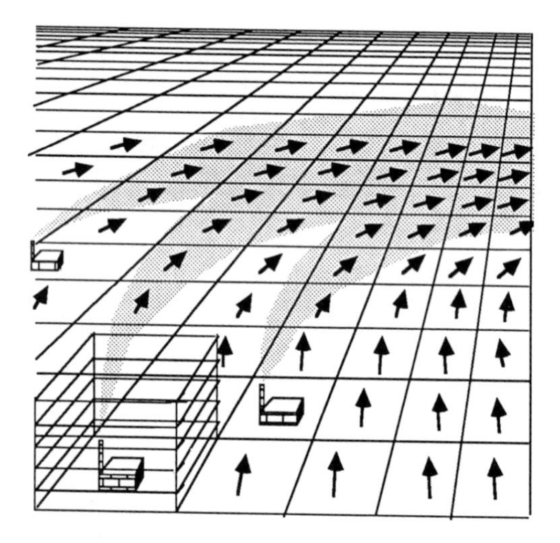

Figure 4. A schematic representation of a Eulerian air quality model. In contrast to the Lagrangian models, more detail can be included in this framework about the vertical and horizontal distribution of winds, as well as more detailed chemical calculations.

There is a third type of model, the *Eulerian model*, illustrated in Figure 4. A model of this type is being developed at the National Center for Atmospheric Research (NCAR) with the support of the EPA [2,3]. Eulerian models has several advantages and some disadvantages. The advantages are that fewer simplifications need to be made about spatial averaging. With less spatial averaging more detailed chemical and physical processes can be included in the model. However, there are still considerable debate about how to best describe those processes and, with the increasing complexity, a growing need to address the influence of uncertainties on results. The main disadvantage of this type of model is expense, both in terms of computer time and the time and labor necessary to interpret the results.

Because of the time and money constraints imposed by complex models, they will be used initially to analyze only a limited number of case studies. Questions arise then such as: can we aggregate a few three-day simulations of wet deposition and dry deposition into an estimate of a long-term pattern of deposition? and will the aggregated cases represent source-receptor relationships accurately?

It may be difficult to take limited case studies, aggregate them to an estimate of longer-term deposition patterns, and come up with a long-term statement about source-receptor relationships. In fact, in the process of aggregation, we may lose the accuracy that was gained by using a detailed model. The research on just how much you lose or gain is still being debated.

It is difficult even to estimate source-receptor relationships for a single case study. As part of the NCAR project an innovative approach, called a carrier frequency, is being used. Each source is given a different frequency, like a radio wave. The concentrations at the grid points are then evaluated using spectral methods to identify and quantify the contribution from each source. This has been done in a preliminary analysis, and holds some promise for application to the whole model.

A second approach for deducing source-receptor relationships from a complex Eulerian model is more direct. You run the model first with a set of control conditions, presumably as close to reality as possible. Next you reduce emissions in some source region of interest and subtract the results from the controlled case. The difference is a direct comparison of the influence of source-receptor relationships for that time period, under those specific environmental conditions

MODEL UNCERTAINTIES

There are a number of uncertainties associated with computer simulation of source-receptor relationships regardless of model type. First, there is a great deal of uncertainty in estimating the transport of pollutants. Second, the parameterization of transformation and deposition rates suffer from lack of reliable data against which to test these components. Third, the transformation and/or deposition processes may not be linear, which raises the question of how do we use source-receptor relationships for a non-linear system? Fourth, there is a great deal of meteorological variability from year to year, and you have to define what that variability is apt to be in order to understand how robust source-receptor estimates are. And finally, ultimately, these matrices cannot be tested. I am unaware of any tracer experiments which have been conducted or planned which will allow us to test the transfer matrix theory over a long time period.

Transport Uncertainties

The uncertainties associated with transport result from our lack of information about the vertical structure of pollutants (and hence which layer is transporting the material) and the coarseness of available non-surface based wind measurements. What we find when we do look at transport in different layers of the atmosphere is that there is a great deal of *shear*, that is, the wind can blow from one direction in one layer of the atmosphere and from a different direction in another layer. Figure 5 illustrates the variations in estimated transport (and hence source-receptor relationships) which occur when assumptions must be made concerning the vertical distribution of a pollutant. If the pollutant is well mixed, as is represented in Figure 5, then the use of a mean wind for a "mixed-layer" of the atmosphere is probably reasonable. The variations in wind flow with height will add to the dispersion of the material, but the estimated mean flow would represent the movement of the pollutant mass. If, in Figure 6, for example, the pollutants are heavily concentrated near the surface, then the used of a layer-averaged wind will misrepresent the transport of the pollutant.

Surface winds alone cannot be used to describe the movement of pollutants since we do know that the material is often contained in layers above the surface. Unfortunately, the only data routinely available is from the network shown in Figure 7 of upper-air stations in the United States and Canada. The distance between those sites is roughly 400-500 kilometers. The data is only taken once every twelve hours, which makes it difficult to interpret what the winds are over, say, Detroit in the afternoon when the nearest available measurement was available from Flint, Michigan (100 km to the north) at 7:00 a.m. EST.

We can test how much error should be expected. Kahl [4], using data from the Cross-Appalachian Tracer Experiment (CAPTEX), has evaluated a number of interpolation techniques ranging from relatively simple to fairly complex. He showed that, even in a quiescent meteorological period, the mean absolute errors due to spatial interpolation to be expected with the existing network of measurements is on the order of three meters per second. Errors of similar size were found for interpolating winds to times between measurements.

Using the "normally" available network of measurements Kahl calculated the trajectory of an air parcel leaving Dayton, Ohio, during one of the CAPTEX tracer release periods. He also computed a trajectory leaving the same site at the same time but using available upper-air data from an expanded network of sites measuring every six hours. The solid lines shown in Figure 8 represent his estimates of where that release would have gone if we used only the existing National Weather Service data; the dotted lines represent where we think that material would have gone if we used that plus those EPRI sites. You can see that you get different answers, considerably different answers depending on how much information you have.

Using a technique called the Trajectory of Errors [5], you can estimate not only the trajectory path but also the probability of that trajectory being elsewhere. Figure 9 shows a solid line indicating the mean movement of air away from Dayton, Ohio, during another tracer release. The dashed lines indicate his estimated confidence levels. The first scale line away from the white line is 25 percent, the next is 50%. Taking those lines, it means the trajectory had only a 50% chance of being between the second line on the top and the

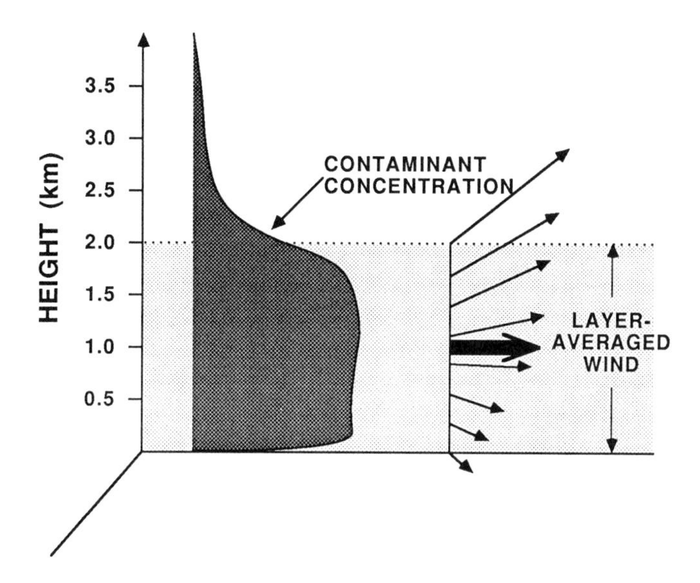

Figure 5. A schematic description of the distribution of winds and pollutants through the "mixed-layer" of the atmosphere when the pollutants are fairly well mixed.

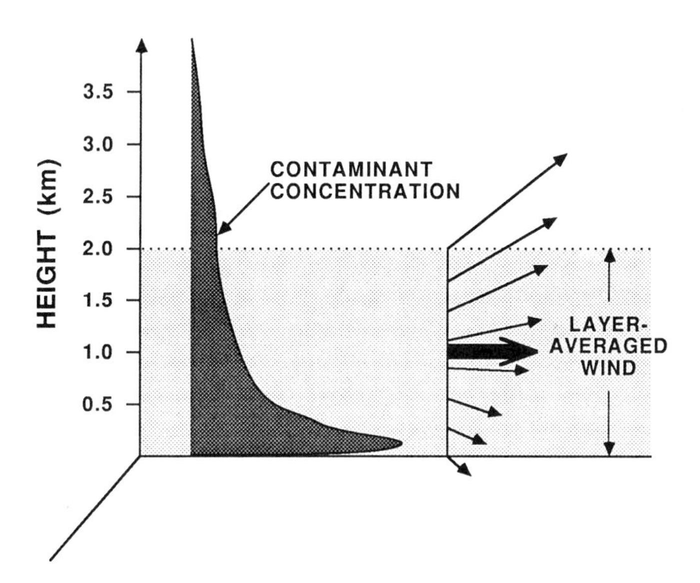

Figure 6. A schematic description of the distribution of winds and pollutants through the "mixed-layer" of the atmosphere when the pollutants are poorly mixed and being advected by winds closer to the surface.

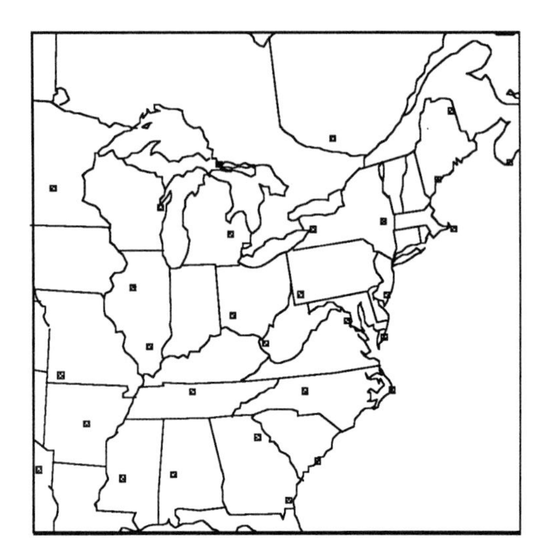

Figure 7. Location of the rawinsonde measurement stations in the eastern United States and Canada. The separation between stations is roughly 400 km and measurements are made twice daily (7:00 a.m. and 7:00 p.m., E.S.T.).

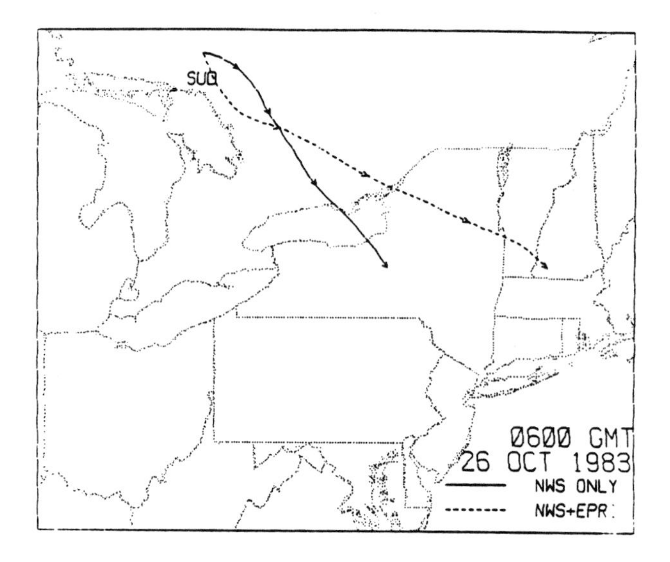

Figure 8. Mixed-layer trajectories originating at Sudbury, Ontario at 1:00 a.m. E.S.T. on October 26, 1983. The solid and dashed lines indicate "normal" data availability and an "enhanced" data set consisting of additional sites with twice the frequency, respectively (from Kahl [4]).

second line on the bottom. This then represents the level of confidence we can have using the existing upper-air network of measurements for air pollution transport studies.

Parameterization Uncertainties

The next question arises from our choice of parameterization schemes used to simplify complex processes. To exemplify this uncertainty, global sensitivity analysis [6] was performed on the Samson and Small [1] model to identify the sensitivity of source-receptor relationships to uncertainties in that model's parameterization. The parameters included the transformation rate for sulfur dioxide and sulfate; the height of the mixing layer; the emissions rate; the amount of precipitation that fell at the receptor; the amount of precipitation that fell upwind of the receptor, and the assumed rate of dispersion. The results [7] showed that the prediction of absolute concentration of sulfur dioxide depends very strongly on transformation rate. Likewise, the predicted concentrations of sulfate are also very strongly linked to our estimations of transformation rate.

Varying the transformation rate, however, had little influence on calculated source-receptor relationships using this model for this site. With the half and double the presumed base transformation rate, the simulated source-receptor relationships did not change appreciably. In fact, it was found that if we took almost any parameter in the model and varied it, we're not going to affect the source-receptor relationships very much at all, with the exception of transport. At least with this model the source-receptor relationships could only be significantly affected through variation in transport patterns.

Uncertainties Due to Non-Linearity

The question of uncertainty in source-receptor relationships due to nonlinearities is beyond the scope of the present paper. If, for example, the rate of production of sulfate is largely due to a reaction between dissolved SO_2 and an oxidant, then the component in lower concentration could control the overall rate of conversion. Moreover, if the oxidant were the component in lower concentration then there would be a limit on how much sulfate could be produced. Thus, raising concentrations of SO_2 would not necessarily increase the concentration of sulfate in precipitation. Unfortunately, the sensitivity of source-receptor relationships to such non-linearities has not yet been evaluated.

Uncertainties Due to Meteorological Variability

We know that meteorology alone has the potential to vary deposition from year to year. Estimating wet sulfur deposition at Hubbard Brook over a six-year period using our receptor-oriented Lagrangian model a range on the order of 6 kilograms per hectare of sulfur was simulated due to variations in meteorological conditions alone. These variations will be due to changes in wind flow associated with precipitation events from year-to-year, changes in precipitation amount at the receptor, and changes in the timing and location of precipitation upwind of the receptor. Since these natural variations in meteorology would be expected to vary the deposition at the receptor, should we not expect source-receptor relationships to also be influenced from year-to-year?

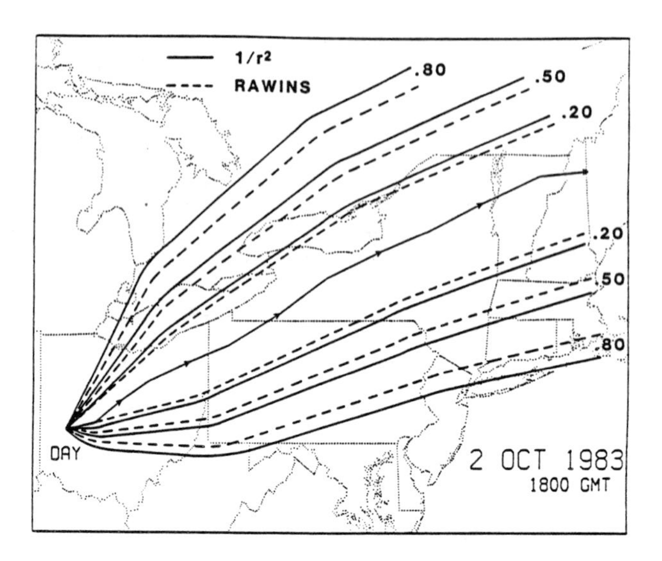

Figure 9. Probable error limits for a mixed-layer trajectory based on interpolation of the current rawinsonde measurement network. The solid lines and dashed represent data interpolation using a very simple scheme ($1/r^2$) and a relatively sophisticated algorithm (RAWINS). Arrows are drawn at 6 hour intervals (from Kahl [4]).

Figure 10 shows the simulated contributions of upwind States and Provinces to sulfur wet deposition at Whiteface Mountain. Our model estimated that there was a range of source-receptor relationships for that site depending on which year chosen. While the largest contribution for most years was simulated to be from sources in the State of Ohio, there were some years when Pennsylvania contributed the most to sulfur deposition at Whiteface according to our model. Thus we have to be cognizant of the fact that source-receptor relationships will change from year to year. If we base our analysis on one year's data of meteorology, we can bias the results.

Testing Simulated Source-Receptor Relationships

Finally, as stated earlier, there is no available artificial tracer experiment planned which will provide the necessary data base against which to test long-term source-receptor relationship estimates. However, in addition to artificial tracers, new results show promise for the use of elemental tracers of opportunity to deduce transport patterns. While this work is in its infancy, the lack of alternative data for testing source-receptor estimates makes this a viable candidate for testing in the future.

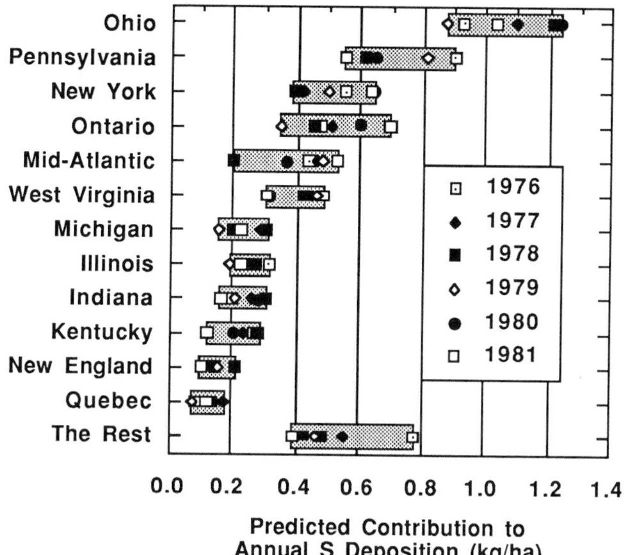

Figure 10. The range of simulated contributions from emissions in various States and Provinces to sulfur wet deposition at Whiteface Mountain, NY during 1980 (from Samson et al. [7]).

SUMMARY

In terms of the transport, we know that even during relatively undisturbed weather conditions, trajectories have a 50 percent chance of exceeding horizontal displacement errors on the order of 250 kilometers after traveling for only about 24 hours. So there is uncertainty in transport. In terms of source-receptor relationships, that is a key. If you can define the transport right, the chemistry apparently becomes much less critical, at least over an annual estimate.

Second, the matrices do not appear to be very sensitive to the transformation or deposition parameters, at least when using a simple model. Third, the sensitivity of the matrices to nonlinearity has not been evaluated, and it should be. Fourth, the simulated source-receptor matrix elements can vary by as much as 70 percent over the six-year period, though generally the variation is less. Finally, insufficient data now exists with which to test the simulated source-receptor matrices. Further, it is unlikely that this data will become usable in the near future. It is possible that these simulations may be tested eventually through the use of empirical data employing elemental tracers of opportunity.

REFERENCES

1. P.J. Samson and M.J. Small in: Modeling of Total Acid Precipitation Impacts, J.L. Schnoor, ed. (Butterworth, Boston 1984).

2. NCAR, The NCAR Eulerian Regional Acid Deposition Model (National Center for Atmospheric Research, NCAR/TN-256+STR, Boulder, CO 1985).

3. NCAR, Preliminary Evaluation Studies with the Regional Acid Deposition Model (National Center for Atmospheric Research, NCAR/TN-265+STR, Boulder, CO 1986).

4. J.D. Kahl, Interpolation of Synoptic-Scale Winds: Implications for Boundary Layer Trajectory Accuracy (Ph.D. Thesis, Univ. Michigan, University Microfilms, Ann Arbor, MI 1987).

5. J.D. Kahl and P.J. Samson, J. Climate Appl. Meteor. 25, 1816 (1986).

6. G.J. McRae, J.W. Tilden, and J.H. Seinfeld, Computers and Chem. Eng. 6, 15 (1982).

7. P.J. Samson, M. Fernau, and P. Allison, Water Air Soil Pollut. 30, 801 (1986).

Non-Modeling Approaches to the Determination of Source–Receptor Relationships

Lester Machta
Air Resources Laboratory
8060 13th Street, Silver Spring
MD 20910

ABSTRACT

After a brief review of ways in which a source-receptor relationship may be obtained, two non-modeling approaches will be described. The first uses the observed decreasing time trend of sulfur dioxide emissions in eastern U.S. from about 1977-1982 and its possible correlations with observed changes in precipitation sulfate concentration or deposition. In the second, hypothetical approach, the possibility of deliberately modulating emissions of sulfur dioxide to see what kind of corresponding changes in precipitation concentrations of sulfates accompanies the emission changes is examined. Problems accompanying each of these two approaches are described.

INTRODUCTION

There are several ways to obtain the source-receptor relationship or the fraction of the pollutant emission from one area which appears in the air or is deposited in the same or another region. Logically, one ought to pick the best estimates of the physics and chemistry of the atmosphere to formulate the transport, dispersion and deposition of a pollutant. Such a physical model can then be applied to any source-receptor combination, to any amount of pollutant emission, as well as the emission or presence of other interacting atmospheric chemicals, and to any time period for which weather conditions can be specified.

The present paper, however, looks at other ways in which one might obtain a source-receptor relationship. Thus, one rational approach might involve a tracer technique in which an identifiable isotope of sulfur or oxygen, for the case of studying the fate of sulfur dioxide, is released in an area and measured in another area. Presumably there should be no or negligible amounts of the isotopes naturally present so the specially introduced tracer can be specifically identified yet the sulfur dioxide so released will behave just like ordinary $^{32}SO_2$. There are unfortunately several difficulties with this idea. The two main stable sulfur isotopes ^{34}S and ^{36}S are abundant in the atmosphere so that relatively large amounts of their isotopes in the form of SO_2 are needed to overcome the natural but variable background. Furthermore, their production is also relatively costly so that to conduct experiments to distances of concern, such as 1,000 km, would involve large expenditures of money. The oxygen isotopes are less suitable than those of sulfur because they can exchange during chemical

Published 1988 by Elsevier Science Publishing Company, Inc.
Acid Rain: The Relationship between Sources and Receptors
James C. White, Editor

reactions. A radioactive isotope of sulfur, ^{35}S, needs to be emitted in much smaller amounts since its cosmic ray produced background is low. An experiment using radiosulfur is technically feasible and may even be able to be conducted within the health standards established by many countries. But, the likely adverse public reaction to the introduction of radioactivity to conduct research on another pollutant has so far thwarted its serious consideration.

Other tracers than sulfur isotopes have also been suggested even though they may not behave chemically or physically similar to SO_2. These other tracers would rather be used as markers of atmospheric motions and together with an elaborate emissions and monitoring program might be interpretable in terms of a source-receptor relationship. This, in general terms, is the basis for the EPRI MATEX study. During peer review of this suggestion, it was felt that it had too small a likelihood for success, considering its cost, to be justified.

The two additional suggestions for the determination of a source-receptor relation to be discussed in this paper involve modulation of the source of sulfur dioxide and correlating this variation with changes in precipitation concentration or wet deposition of sulfates. It could, of course, be applied to NO_x emissions and subsequent measurements elsewhere of nitrate concentrations or deposition, etc.

DELIBERATE MODULATION OF SO_2 SOURCES

The most direct way of finding out whether a reduction in the deposition of sulfates will occur in the Adirondack Mountain area from a reduction in the emissions of SO_2 in the midwestern U.S. is to try it. A scientific analysis of such a possibility indicates that detection of the change in concentration in precipitation or deposition requires both an extended period of emission reduction and a fairly large amount of reduction. The cost associated with such a protracted, large reduction is estimated in the billions of dollars. The reason for such large amounts of sulfur dioxide reduction for detection of its consequences is the very high natural variability in the concentration of sulfates in rain and in the amount of precipitation. If this large variability of concentration of sulfates in precipitation were to be significantly reduced, it might then be possible to detect changes in the receptor area from much smaller emission reductions at much lower cost.

There are two currently known and readily predictable reasons for the high variability of the concentrations: first, air passes over regions with very diverse source amounts of sulfur dioxide such as the difference between oceans and heavily industrialized regions and second, there is a known decrease of concentration with the amount of precipitation which may in part result from the fact that the early part of a rainy period generally has higher concentrations than the later period. In this section, the decrease in standard deviation of concentration variability divided by the mean, a measure of sulfate in rain events (really rainy days), will be compared before and after adjustments are made for the amounts of precipitation and for the regions over which the air moved to reach three sampling stations in New England.

An area centered in eastern Ohio, a region of currently high sulfur dioxide emissions was selected as the hypothetical target area in which a source modulation might occur. The receptors were taken from three of the utility UAPSP stations which report daily precipitation concentrations: Big Moose, NY; Winterport, ME; and Underhill, VT. The data were selected at these stations for three years (1982-84) during three summer months: June, July and August. Since the concentrations of sulfate ions in precipitation more closely approximate a log-normal rather than a Gaussian distribution, all of the subsequent calculations use the logarithms of concentration (millimoles per milliliter of precipitation). The standard deviation divided by the mean of the logarithm of concentration for all cases of precipitation, some 271 cases, was 32%.

When only those cases (96) were chosen in which the back trajectories from the three stations which passed over the target area, the same statistics reduced to only 16%, one half of the original measure of the variability. The mean value of the logarithm of concentration increased by about 16% confirming a larger source in eastern Ohio than, on average, from other directions backwards from the three receptors.

The adjustment for the amount of precipitation, assuming a log-log linear relationship (concentrations are generally greater when amounts of precipitation are smaller), was disappointingly small for this body of data involved. Using the cases in which trajectories passed over eastern Ohio and also adjusting for the amount of precipitation reduced the standard deviation divided by the mean from 16% to only 15% (or about a 10% reduction in variance). At other National Trends Network locations and times the reduction in variance was greater than 10%, perhaps in the range of 20 to 50%. Incidentally, if one examines the reduction in variability of concentrations of other ions not connected particularly with the industrialized eastern Ohio, one finds that there is a slight variability reduction, 25% for calcium, but no decrease in variability for sodium when cases of selected trajectories over Ohio are compared to the entire body of data.

A major concern with the above statistical treatment of data deals with the independence of the receptor measurements. Thus, to take an extreme example, suppose that all the receptor sites (and there might be very many) had only a few cases of back trajectories passing over Ohio and all happened to originate from an emission from the same few days. This situation would not be truly sampling a large enough number of emission cases to be representative of the emissions.

Nevertheless, the little work that has been undertaken so far suggests that a significant reduction in variability can be achieved. This, in practical terms, means that a shorter and smaller decrease in concentration might suffice to confirm or reject the expected benefit from an emission reduction. If such a modulation experiment were to be undertaken, the selection of cases in which the air passed over the target region could be made more certain by the use of inert tracers released from the target area to the observed receptor area which would substitute or augment the meteorological trajectories on which the above statistics were based. But, most important, the cost analysis to reduce emissions would have to be reexamined; reducing the cost from two billion to only one billion dollars still keeps it out of reach.

PAST VARIABILITY IN SULFUR DIOXIDE EMISSIONS

In the U.S., emissions of sulfur dioxide have tended to decrease annually since the late 1960's or early 1970's. Not all parts of the U.S. have shown the same trend; for example, the southern U.S. has tended to exhibit a smaller change than the industrialized northeast. The decrease in the northeast and adjacent area has amounted to a total decrease between 1977 and 1982 of about 20%.

Although the trends depend somewhat on the treatment of the data, such as smoothing of data, the average 10-year percentage decrease of the annual average sulfate concentration in precipitation between 1964 and 1984 is greater than 2% per year at Hubbard Brook, NH. This is the station with the longest continuous, quality record. Most other stations used in trend analysis generally began in 1977 or later. A U.S. Geological Survey analysis of 13 or 14 stations of the National Trends Network over the entire contiguous U.S. exhibited a yearly decrease of 8 to 10% while six to eight stations east of the Mississippi River, other than Hubbard Brook, showed an average annual decrease of 6 to 8%. Not all of the stations, individually, showed statistically significant decreases. But many stations exhibited a larger decrease than was observed in the emissions of sulfur dioxide.

The above comparison between a decrease in emission of SO_2 and associated decrease in concentration of sulfate in precipitation, even if not one-to-one, strongly suggests that there will, in fact, be a benefit in lower concentrations from a reduction in emissions over a wide area. However, there is evidence which argues for caution in reaching such a conclusion.

An analysis of sulfate in precipitation at 5 stations from the MAP3S network operated between about 1977 and 1984 (two stations near the sea coast were left out because they need sea salt sulfate corrections) suggest that, depending on the treatment of data, the decrease will be less than 1% per year to less than 2% per year, much less than reported by the more numerous National Trends Network stations.

But perhaps the more puzzling aspect of the wet chemical monitoring data in all the networks is the fact that some of the other ions collected in rainwater such as sodium and calcium exhibit even more dramatic decreases in concentration during the same interval and at the same stations as does the sulfate. The early sodium ion concentrations in the MAP3S network are suspect. It is highly unlikely that the source of the sodium and calcium changes is from man's activities. This has led the National Academy of Sciences to contend that "...concentration changes from year to year occur concurrently in both anthropogenic and natural constituents and are thought to be more strongly affected by meteorological rather than emission changes..."

Thus it can be argued on the basis of measurements in the late 1970's and early 1980's that an emission control strategy similar to the roll-back of emissions of sulfur dioxide over a large area might result in a decrease in sulfate concentration in precipitation in downwind areas. But there are still sufficient doubts raised by the failure of several good stations to detect the change and, even more significant, because the parallelism of man-made sulfates and several natural substances remains unexplained. These aspects create doubts whether a control strategy of a magnitude and area similar to that which has occurred in the '70s and '80s will truly be effective in reducing concentrations of sulfate in precipitation.

Questions and Answers following Modeling and Data Analysis Session

Thursday, December 4, 1986

Q:

My question relates to remarks of a very good scientist who said about three years ago that in all the models the deposition of a given amount of sulfur would probably vary by the ratio of the square of the distance, meaning that if you had a source in Pennsylvania that was emitting a certain number of tons of sulfur, and it was 300 miles away from the Adirondacks, that source would be at least four times more significant than a source 600 miles away. He claimed that models assumed distance to be insignificant. Could you comment on that?

A: (Samson)

Dealing with distance, "all other things being equal" is a quite a caveat, but if things being nonreactive were being emitted into the atmosphere with winds blowing from all directions equally, and all other meteorology being the same, the relationship you stated might be true. Which is not a very strong statement. What we do with our models is look at which way the winds <u>do</u> come from. Of course we expect sources further away to have potentially less impact if they are nonreactive; but if they are reactive, it is not immediately obvious that a source farther away will have less impact than those closer in. Remember that you have sulfur dioxide being converted to sulfate, and that does take some time, so many things will happen in between. Because of the meteorology, because trajectories vary over the course of the years, sources closer probably have a larger chance of contributing. Taking the complicated chemistry, as these models do, then hopefully we get a little bit better estimate than the type of approach your scientist suggested. Basically it is true, but once you add chemistry, there are many differences.

Q:

If you could just comment generally on how valid the figures are that were used by the State of New York, perhaps from your model. I've seen, at several conferences, figures of the contributions of various states to deposition at Whiteface Mountain. Indiana's is maybe three percent or five percent. Ohio is an in-between case. But one statement you made seemed to suggest that those figures were pretty reliable; you said they don't depend very much on transformation rate. Another statement you made seemed to suggest they're quite unreliable, and we shouldn't be basing any legislation on them. Wouldn't that be unreasonable?

A: (Samson)

Would it be reasonable if they listen to the scientists, and take the best available evidence, promulgate the legislation based on that? While we don't claim these models are accurate by any means, we think they are reasonable estimates, the best available knowledge, for how air gets from a source to a receptor.

Comment:

A comment, since you used the Hubbard Brook data. There is a huge difference between the length of the records at other sites and those at Hubbard Brook. I've overlayed the MAP3S data on top of the Hubbard Brook record and found the MAP3S for 1983-84 almost exactly the same as the Hubbard Brook data. They are not any different during that period of time.

Q:

I'd like to ask Dr. Machta about future field experiments. I think at the most recent Atmospheric Task Group review there was a lot of big plans presented on field experiments. What has happened with that, or any other field experiments?

A: (Machta)

Right now we are trying to formulate plans for a validation of the Eulerian model. These plans are not yet to the stage where one can describe when or exactly how they will be conducted. I am involved in one experiment in which we are going to release an inert tracer during the CAPTEX experiment in Montana, to try to validate the transport component of the transport dispersion and deposition model. That will go off in January to March of this coming year.

Source–Receptor Relationships: The Canadian Experience

James W. S. Young
Atmospheric Environment Service
4905 Dufferin Street, Downsview
Ontario, M3H 5T4

ABSTRACT

In 1982 the Canadian government decided to base its control strategy for acid rain on science. This paper will review the scientific approach taken (optimization) in the development of its "selective reduction" strategy, give some examples of the application of this science to support the strategy and talk about the conclusions reached in 1982. The paper ends with a review of these conclusions based on advances in the science between 1982 and 1986.

HISTORICAL PERSPECTIVE

On 5 August 1980, the governments of the United States and Canada signed a Memorandum of Intent "to develop a bilateral agreement on transboundary air pollution including the already serious problem of acid rain". The long range transport of acidic pollutants in the atmosphere and their subsequent deposition to sensitive receiving surfaces such as fresh water bodies and soil was identified as a matter of increasing concern [1]. One of the aims of the Memorandum of Intent was to further the development of effective emission control programs and offer measures to combat transboundary air pollution.

To provide a suitable technical and scientific basis for the formulation of such an agreement, bilateral work groups were established to prepare scientific reports on specific aspects of transboundary pollution. The Atmospheric Sciences and Analysis Work Group (Work Group 2) attempted to define the applicability and limitations of eight long-range transport (LRTAP) models for assessing the result of changes in annual S emissions rates in eastern N America [2]. These LRT models used S emission and meteorological data to predict fields of concentration and deposition of S compounds that had been implicated in the process of acidification of the receiving environment. It will be shown later that these models can be useful tools in developing emission control strategies.

S emission abatement to reduce acid deposition is expensive. It is important that deposition be reduced to acceptable values while, at the same time, either the amount of S removed or the cost of doing so be kept as small as possible.

In 1982, Canadian Federal and Provincial governments agreed that the Canadian environment would be protected from acid rain damage if wet sulphate deposition were reduced to a level of 20 kilograms per hectare per year (18 lbs/acre). This level was defined as a result of cooperative scientific research later reported under the Memorandum of Intent [3]. Many areas in Canada were receiving and continue to receive up to twice that amount of sulphur.

In 1982 the Canadian government decided to base its control strategy for acid rain on science. Science had given us an understanding of the nature, causes and solutions to the acid rain problem. Science had

Published 1988 by Elsevier Science Publishing Company, Inc.
Acid Rain: The Relationship between Sources and Receptors
James C. White, Editor

identified trends in acid deposition and an environmental objective to protect moderately sensitive lakes and rivers from acid rain damage. It was felt that science could also point out where emissions should be reduced to achieve the objective.

APPROACH

The approach used is detailed in Young and Shaw [4] and will be summarized here for completeness.

Atmospheric Transfer Matrices

For the purpose of estimating changes in concentrations and depositions due to emission changes, the output of the LRT models can be conveniently expressed by an atmospheric "transfer matrix". The matrix formulation assumes that the deposition at a receptor location is the sum of partial contributions, each of which is proportional to the emissions from a source or group of sources. Thus, the deposition D_A at receptor A would be of the form:

$$D_A = T_{A1} \cdot E_1 + T_{A2} \cdot E_2 + \ldots + T_{AJ} \cdot E_J + \ldots + T_{AN} E_N$$

where E_J is the emission rate in source region J and T_{AJ} is a coefficient of proportionality connecting the source region with the receptor A. The array of the coefficients connecting deposition at a receptor with a unit of emission in a source region constitutes a unit transfer matrix and indicates the strength of the atmospheric linkage between source and receptor. The three parts of Table I (A, B and C) are a hypothetical example of the relationship among emission rates, a 3 x 3 unit transfer matrix and the resulting absolute deposition values. The unit transfer matrix (Part B) indicates that there is a strong atmospheric transfer between Source 3 and Receptor 3 (4.0 kg ha^{-1} of S is deposited at Receptor 3 per teragram S per year emitted at Source 3). In contrast, there is no atmospheric transfer between Source 3 and Receptor 1, because the matrix element is zero.

TABLE I. Hypothetical example of relationship among emissions in three areas (Part A), atmospheric unit transfer matrix (Part B), and the absolute deposition (Part C) at three receptors.

A		B			C			
		Unit transfer matrix (kg ha^{-1} deposition per $T_g y^{-1}$ emission)			Absolute deposition matrix (kg ha^{-1} y^{-1})			
		Receptor			Receptor			
Source No.	Emission rate	1	2	3	1	2	3	Total
1	24	0.9	0.5	0.7	21.6	12.0	16.8	49.4
2	13	0.1	0.6	0.0	1.3	7.8	0.0	9.1
3	2	0.0	0.8	4.0	0.0	1.6	8.0	9.6
Total	39				22.9	21.4	24.8	

Multiplying the emission rate column in Part A of Table I by the appropriate elements of the unit transfer matrix (Part B) will give the absolute deposition contribution from each source at each receptor (Part C) which may be summed to obtain the total deposition.

The horizontal totalling in Part C indicates the relative overall impact of each source over the three receptors (the higher the total, the greater the impact). In the example in Part C, Source 1 has the greatest relative impact on the three receptors. The horizontal summing may be carried out over a subset of the receptors; for example, over Receptors 1 and 3 only.

The MOI modellers decided that the emission regions to be used in the development of a transfer matrix should be resolved spatially on a province/subprovince and state/multi-state level with 15 source regions in Canada and 25 in the U.S.A., as shown in Fig. 1. Unit transfer matrices have been produced linking emissions in the 40 source regions with deposition at nine receptor points (shown as number circles in Fig. 1), where ecosystems are sensitive to acid deposition.

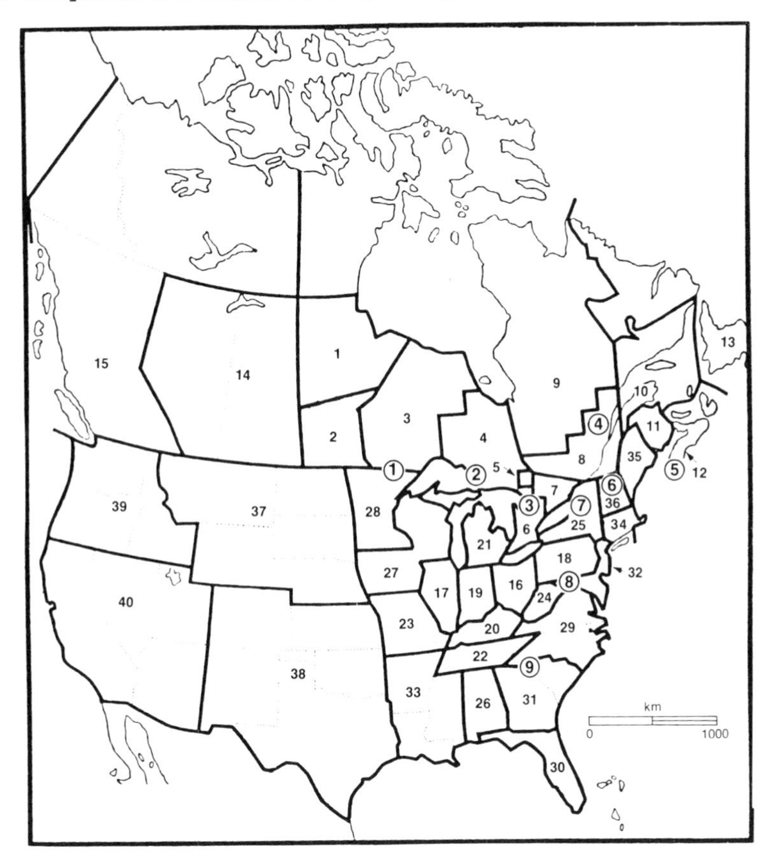

FIG. 1. The 40 source regions and the 9 sensitive receptor points as designated by the modellers working under the United States-Canada Memorandum of Intent on Transboundary Air Pollution [2].

Because transfer matrices display the results of model calculations, they suffer from any limitation in the models and in the input data. Some additional limitations are:

(a) The aggregation of emission sources within a source region and the averaging of emissions over space and time means that additional emissions from a new source may not behave exactly as indicated by the transfer matrix value.

(b) The matrices do not include the effects of other pollutants which might alter the S chemistry. Shaw and Young [5] and Oppenheimer [6] suggest, however, that this non-linearity might produce at most a 10% error when applied to predictions of total S.

(c) Matrices are not suitable for assessing the deposition within 100-200 km of a given source, because of the coarse spatial resolution of the meteorological data used in the models, and the relatively coarse division of N America into only 40 source areas (Fig. 1).

(d) For control strategy purposes, the limited number of receptor points (in this case, nine) may not include all of the environmentally sensitive areas that one wishes to protect.

In spite of these limitations, Canada decided that the methodology described below was appropriate for general guidance in defining emission reductions in relatively large geographical areas.

Optimization Methodologies to Minimize Emission Reductions

The four optimization schemes described below make use of either the unit transfer matrix or the absolute deposition matrix for ranking the source areas. The ranking is based on deposition per unit emission or deposition in an absolute sense. The approach is to reduce emissions first in the source area which is deemed to have the largest effect on a single receptor or a group of receptors. Three methods were aimed at reducing deposition at a single receptor point. For example, from Table IC, one could choose to protect receptor area 3 (24.8 units of deposition). Protecting receptor 3 will also help protect other receptors. One method (S1) uses the unit matrix (Table IB) to rank sources and starts with the largest transfer element (meteorological linkage). For example, to protect receptor 3 emission reductions would first be applied to source 3 until either the deposition is reduced to 20 kg/ha/yr or the emissions in source region 3 reach a maximum allowable level (based on technology, politics or other considerations). Then one would move on to source 1. Method S2 is similar to S1 but uses the absolute deposition matrix (Table IC) to rank sources. Here one would reduce source 1 first to protect receptor 3. In all cases the ranking is re-assessed after each increment of emission reduction. A third method (S3) did not reduce emissions sequentially but rather simultaneously. Each source's share of the emission reduction was proportional to its contribution to the deposition. The fourth method (Y1) was based on a grouping of receptors using the absolute transfer matrix as a field (Table IC). Examining the horizontal totals (relative impact of one source over many receptors) shows us that the impact of source 1 is greatest. A reduction is then applied to source 1 first and follows the same general procedure above except that all receptor areas are examined before an emission reductions step. Let us now move on to a real example in North America.

EXAMPLE OF APPLICATION

Model Used

The scientific recommendation to the strategists was based on the wet deposition matrix from the MCARLO Model [7] on the basis of five criteria described in Young [8]. Figure 2 is a plot of the model-predicted wet deposition for 1980 at the nine sensitive receptor areas along with the measured data. The observed data were plotted from left to right such that they form a monotonically decreasing set. This allows one to examine the general spatial characteristics of models. The only model that subjectively performed in the same way was the MCARLO model. This measure plus an examination of the error, the scatter of the error, the ratio test and the random error distribution pointed out the superiority of the MCARLO model.

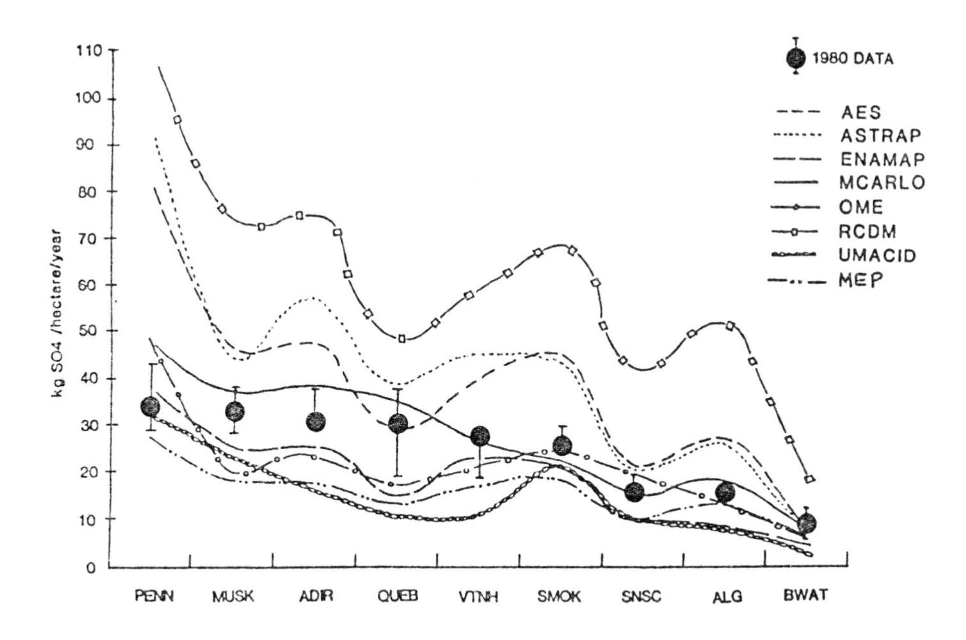

FIG. 2. 1980 Measured and Predicted Wet Deposition at Selected Sensitive Areas.

It was obvious that 100% reduction of emissions in a given source area was impossible and that reasonable constraints based at least upon technological considerations had to be used. For the 15 emission areas in Canada, the Environmental Protection Service of Environment Canada estimated maximum allowable emission reductions based upon the distribution of sources (mainly non-ferrous smelters, fossil fuel burning electrical generating stations and residential and commercial furnaces burning fuel oil) and the then available control technology for each of those sources.

In the U.S., approximately 80% of SO_2 emissions in 1980 came from electrical generation and residential, commercial and industrial boilers [9]. Due to a lack of information, a uniform value of 70% maximum allowable emission reduction in all 25 U.S. emission areas was used as an approximation for the purpose of demonstrating the effect of constraining the emission reductions.

Figure 3 is a map showing the emission reductions according to method Y1. The amount of S that must be removed in N America is larger than that in the theoretical optimum (100% reduction allowed) because emission reductions are being made in source areas that are meteorologically less closely linked to the receptor areas of interest.

It is also interesting to note that the imposition of constraints upon the emission reductions tended to reduce the differences among the estimates produced by the different methods of overall S removed from N American emissions.

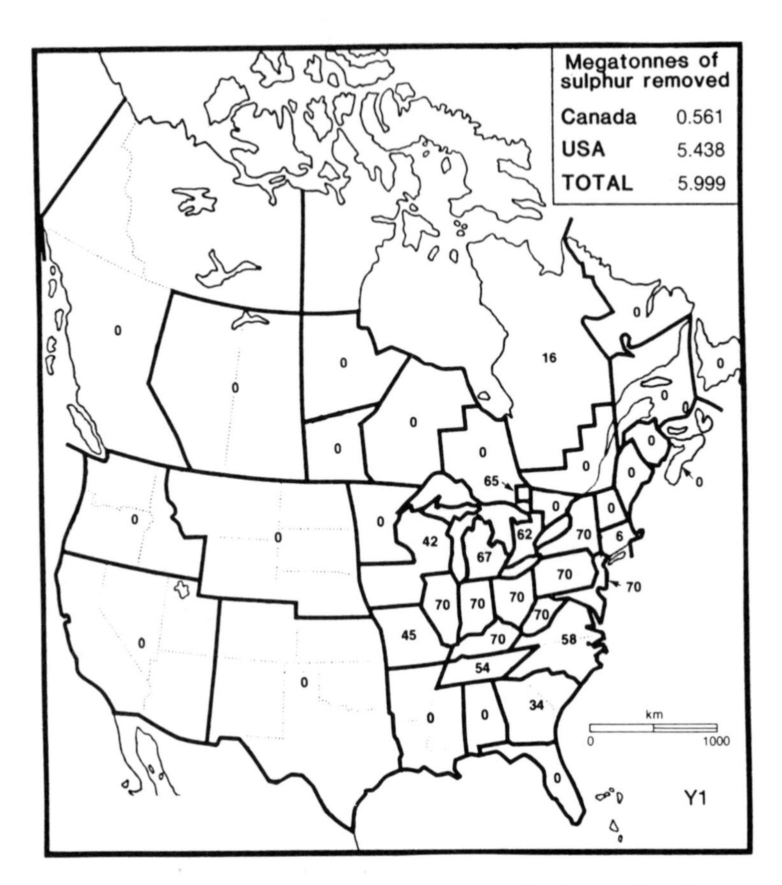

FIG. 3. 'Technological' optimum percentage emission reductions required in the 40 N American source regions to reduce deposition to 20 kg ha^{-1} y^{-1} at the 9 receptor points in Fig. 1 using maximum allowable emission reductions that consider the capabilities of control technology.

If emissions were to be reduced uniformly across all source areas, it has been estimated, using the MCARLO source receptor matrix, that an overall emission reduction of 65% or 9.5 Mt S y^{-1} should be removed from emissions, to meet the 20 kg SO_4 ha^{-1} y^{-1} wet deposition target at all receptor points. This is considerably greater than the value of 6 Mt S removed each year in Fig. 3. At an average cost, conservatively estimated at $1000 t^{-1} S removed, the savings that could be realized by reducing emissions in an optimized instead of a uniform manner were projected to be in the order of $3-4 x 10^9 y^{-1} for all of N America. The optimization suggested savings of about one-third.

CONCLUSION - THEN (1982)

Strategists recognized that the LRT models (or simple climatic models for that matter) (1) do not incorporate everything we know, (2) should not be applied to problems for which they are not designed (control strategy for a single emission source) and (3) can and have been applied to major economic decisions (building a dam).

On the basis of these results and an in-depth examination of all science evidence, Canada adopted a "preliminary cost-effective selective reduction" control strategy. This has led to an agreement by Federal-Provincial Environment Ministers on the apportionment of SO_2 reductions in Canada. This agreement was reached on February 5, 1985 as outlined below.

"The provincial Environment Ministers agree to the following reductions in SO_2 emissions toward achieving the 1994 objectives:

Province	1980 Base Case (tonnes)	Reductions (tonnes)	%	Emission Objectives (tonnes)
Manitoba	738,000	188,000	25	550,000
Ontario	2,194,000	1,164,000	53	1,030,000
Quebec	1,085,000	485,000	45	600,000
New Brunswick	215,000	30,000	14	185,000
Prince Edward Island	6,000	1,000	17	5,000
Nova Scotia	219,000	15,000	7	204,000
Newfoundland	59,000	14,000	24	45,000
TOTAL	4,516,000	1,897,000	42	2,619,000

Recognizing the overall goal is to achieve a deposition level not greater than 20 kilograms per hectare per year and an eastern Canadian emission level of 2.3 million tonnes (50% of 1980 Base Case), the federal and provincial Ministers are committed to pursuing further reductions in sufficient time to achieve this overall goal by 1994."

EXTENSIONS 1982-1986

Year-to-Year Meteorological Variability - Deposition

Observed year-to-year fluctuations in wet S deposition at a given sampling point were calculated from data reported in the Data Summaries of the Canadian Network for Sampling Acid Precipitation (CANSAP), available from the Atmospheric Environment Service, 4905 Dufferin Street, Downsview, Ontario, Canada M3H 5T4, and from data in the U.S. reported in Barrie and Hales [10]. The ratio of the minimum to maximum annual observed wet S deposition at several regionally representative monitoring points in eastern N America (Table II) ranges from 1.2 to 2.2 with an average value of 1.7.

It is useful to compare the observed year-to-year fluctuations with those predicted by a model using meteorology for different years. Unfortunately, few transfer matrices for a given model during different years have yet been produced. Shannon (1984, private communication) has produced transfer elements for the Advanced Statistical Trajectory Regional Air Pollution (ASTRAP) model [11], linking emissions in 11 source regions with deposition in the Adirondack Mountains for 1978, 1980 and 1981. The results are shown in Table III. It can be seen that, for all but one of the source regions, the transfer element varied over less than a factor of 2 (minimum to maximum). The mean value of the ratio of minimum to maximum in Table III is the same as the observed mean ratio in Table II. This is evidence that models may be able to realistically represent year-to-year variability, evidence which increases our confidence in the models.

TABLE II. Ratio of maximum to minimum wet sulphur deposition at regionally representative sampling stations in eastern N America over sampling periods of several years.

Station	Sampling period (y)	Max/Min
Truro, Nova Scotia	5	1.3
Quebec, Quebec	4	2.2
Maniwaki, Quebec	5	1.4
Pickle Lake, Ontario	4	2.2
Mount Forest, Ontario	4	1.7
Atikokan, Ontario	5	1.6
Whiteface, New York	5	1.2
Ithaca, New York	5	1.7
Penn State U., Pennsylvania	5	1.8
	Mean	1.7

More important, however, is the fact that although the absolute values of the transfer elements in Table III fluctuate from year to year due to different meteorological conditions, their ranking is very similar during the 3 years (Table IV). Since the optimization methodology uses the transfer elements for selecting which source regions are to be abated, the results in Table IV suggest that year-to-year meteorological fluctuation may not greatly affect the solution, at least for those source regions causing the most deposition per unit emission.

TABLE III. Source-receptor transfer elements relating wet sulphur deposition (kg S ha^{-1} y^{-1}) in the Adirondacks with emissions (Tg S y^{-1}) in 11 states (Shannon, J.D., personal communication, 1984).

State	1978	1980	1981	Max/Min
Ohio (OH)	1.50	1.20	1.42	1.3
Illinois (IL)	0.88	0.56	0.70	1.6
Pennsylvania (PA)	2.68	1.39	1.89	1.9
Indiana (IN)	1.21	0.75	0.79	1.6
Kentucky (KY)	0.87	0.64	0.63	1.4
Michigan (MI)	1.59	1.20	1.10	1.5
Tennessee (TN)	0.46	0.51	0.50	1.1
Missouri (MO)	0.69	0.41	0.52	1.7
West Virginia (WV)	0.98	1.33	1.64	1.7
New York (NY)	9.64	2.75	3.43	3.5
Florida (FL)	0.01	0.11	0.14	1.2
			Mean	1.7

Year-to-Year Meteorological Variability - Flux

A recent evaluation by Olson (1986, personal communication) examined the flux of U.S. emissions into Canada for the years 1980-1983. Figure 4 shows that, assuming fixed emissions, the flux from the U.S. to Canada reached a maximum in 1982 which was the result of increased wind flow into Canada during 1982. This points out the importance of meteorological variability. Superimposed on this meteorological variability are the emissions variations which decreased in the U.S. from 1980 to 1982 then slightly increased in 1983. Figure 4 also shows that the reduced U.S. emissions produced a slightly decreasing transboundary flux from 1980 to 1982 and then sharply dropped in 1983, in spite of a slight emissions increase, in 1983, in response to a decreased wind flow into Canada.

The effects of annual meteorological variability on transboundary fluxes, between 1980 and 1983, are slightly less than the annual emissions changes (± 20%). From this analysis, Fig. 4 shows a general downward trend in transboundary flux from the U.S. to Canada.

TABLE IV. Ranking of the states in Table III by decreasing magnitude of transfer element.

Ranking	1978	1980	1981
Highest	NY	NY	NY
	PA	PA	PA
	MI	WV	WV
	OH	OH	OH
	IN	MI	MI
	WV	IN	IN
	IL	KY	IL
	KY	IL	KY
	MO	TN	MO
	TN	MO	TN
Lowest	FL	FL	FL

Model Uncertainty

Young and Shaw [4] carried out an experiment to examine the effect of a ± 50% change in individual elements of a transfer matrix. The effect of varying elements of the matrix can have two effects: (1) the choice of which source regions should undergo emission reductions may change from those selected from the original matrix (the wrong source regions could be selected) and (2) the resulting deposition after emission reductions would be different from that estimated using the unaltered matrix.

The experiment showed that, using the original matrix, 78% of the required deposition at a given receptor was achieved by reductions at the six highest ranked source regions. If this choice of reductions in the top six regions were then applied to the randomly altered matrices, they were found to account for 71-86% of the deposition reduction. This leads to the conclusion that the choice of the highest ranking source region is not very sensitive to fluctuations in the matrix elements.

With respect to the effect of matrix element fluctuations on deposition, the experiment showed 60% of the cases were within ± 2 kg S ha^{-1} y^{-1} of the result using the unaltered matrix.

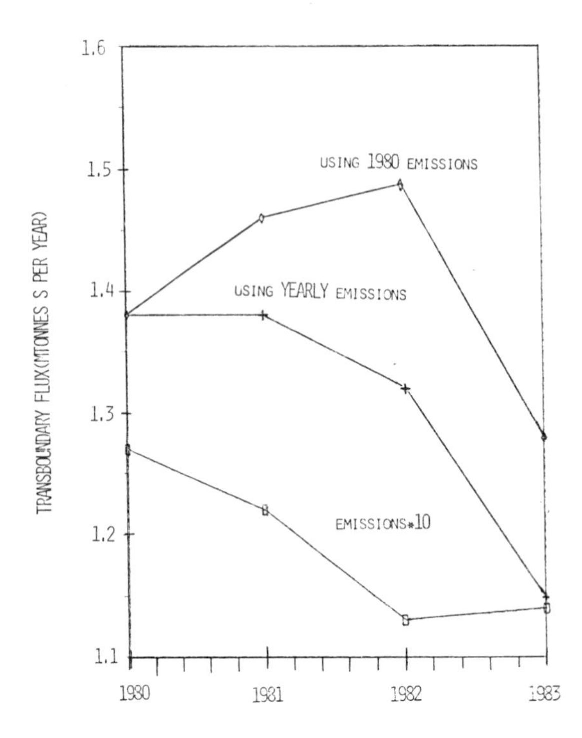

FIG. 4 Transboundary Fluxes 1980-1983 calculated using the AES Model.

Model to Model Variability

To examine the effect of model-to-model variability rather than variability within a given model, optimization runs were carried out using method Y1 and the seven MOI models. To ensure that the optimizations from the seven models were not unduly constrained, the runs were carried out assuming maximum allowable emission reductions of 100%. The minimum, maximum and mean percentage reductions from five models and the estimate from the MCARLO model are shown in Fig. 5 for the 17 source regions requiring the greatest emission reductions, as well as for Canada, the U.S.A. and N America. For each source area, the two outliers (highest and lowest among the seven models) are not plotted in Fig. 5.

The results from the remaining five models best agree for those source regions requiring the greatest emission reductions. Therefore, we have confidence in the model estimates for those source regions which were ranked the highest (and which have the greatest effect upon deposition); our confidence decreases for source regions which are ranked lower.

Furthermore, Fig. 5 indicates that the estimates by the MCARLO model fall, for the most part, within the range of the five models. When they fall outside the range they are lower, indicating that the MCARLO model was fairly conservative and did not overstate the need for emission reductions. For source region 22, the percentage reduction predicted by the MCARLO model is itself an outlier. The reason for this is not clear; obviously the MCARLO model ranks source 22 differently than do the other models.

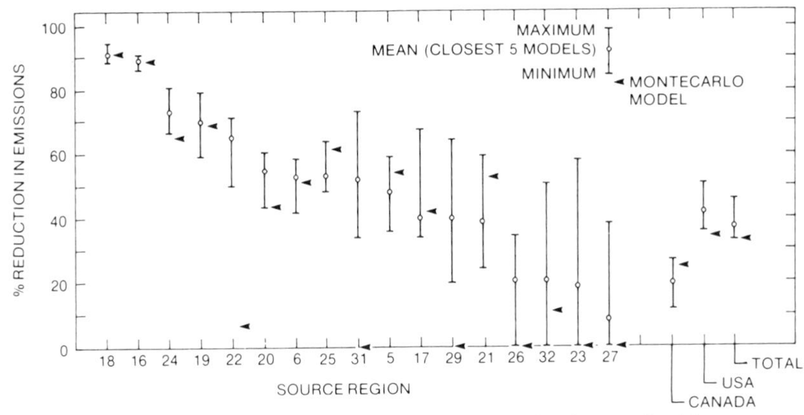

FIG. 5. Range and mean values of sulphur emission reductions estimated by 5 LRT models for the 17 most important source regions and for Canada, the United States and N America. Emission reductions estimated by MCARLO model (arrows) are shown for comparison.

The International Sulphur Deposition Model Evaluation

Recognizing the need to evaluate regional sulphur deposition models on a common basis and to determine whether these models satisfactorily simulate wet sulphur deposition patterns, the U.S. Environmental Protection Service and Environment Canada coordinated the International Sulphur Deposition Model Evaluation (ISDME). This study evaluated 11 models using seasonal wet deposition amounts calculated from screened 1980 precipitation chemistry data. One of the ISDME goals was to establish for the policy analyst the reliability of a model to produce source-receptor relationships. This evaluation investigated whether a model gave the right results for the right reasons. The limited results published to date indicate that at least one of the original 8 MOI models (the MOE model) neither significantly under/overpredicts at approximately the 95% confidence level across more than 20% of the evaluation region of eastern N America nor more than 25% of any of the four subregions [12].

Since the strategy calculated with the MOE model did not differ significantly from the MCARLO model, this result again supports the 1982 decision.

Minimizing Costs

Shaw [13] examined cost data for various control steps for smelters, thermal generating stations and fuel oil desulphurization and optimized emission reductions based on minimizing costs of control. Emission reductions selected by this method cost less than those indicated by an optimization minimizing S removal only. The savings are greater when the costs of the various control steps are spread over a wide range and not clustered about a mean value. In that case, the control costs, especially in the early to middle steps of the control program, may be only a fraction of what they would otherwise be. The curves labelled C in Figure 6A show the results using the cost-optimization. The abscissa shows the wet deposition (and cost) predicted at Receptor 8 (consistently has the greatest wet deposition). For the purpose of comparision, results using method S1 in Ref. [4] is plotted as curve S. In this case the distribution of control costs is fairly uniform among various source steps. If the costs for all control steps are identical curves C and S would coincide.

Shaw [13] also examined the case for Canada alone where the three classes of sources have quite different control costs. Figure 6B gives similar results for reducing deposition at Receptor 3 in Canada using the same methods. As shown in Fig. 6B the cost effective method results in more S removal at each stage of deposition reduction. Fig. 6B also indicates a considerable saving can be realized by including the cost factor in the optimization scheme in the early and middle stages of deposition reduction. This result has been confirmed by Streets et al. [14] and Batterman et al. [15].

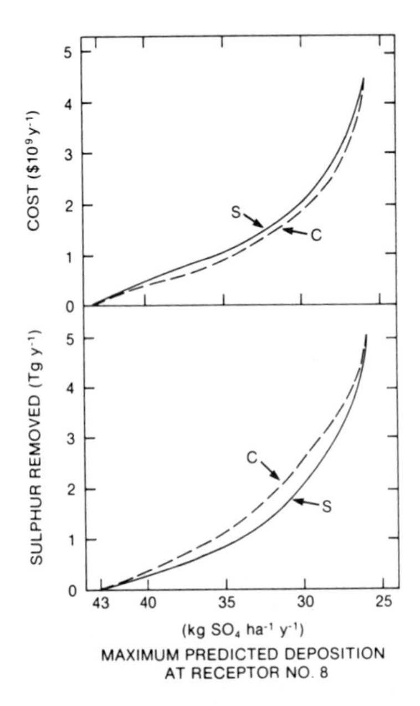

FIG. 6A. Amount of sulphur removed (lower panel) as the wet deposition is reduced at receptor point 8. Curves C are results for the optimization scheme in Shaw [13] which minimizes control costs; curves S are for the scheme described in Young and Shaw [4] which minimizes the amount of sulphur removed.

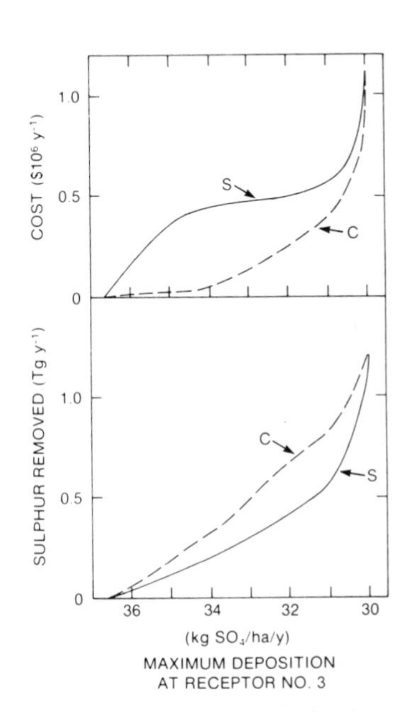

FIG. 6B. As for Fig. 6A but for reducing deposition at receptor point 3 using sulphur removal only from source regions in Canada. (Reproduced with the permission of Shaw [13].)

Other Factors

The model from Ref. [15] also provides a tool for analysis of a wide range of emissions, cost and environmental indicator alternatives. No model to date in this field has attempted to model the attitudes of decision makers such as their behaviour under uncertainty. Using these types of models [4, 13, 14, 15] in an interactive mode could however accommodate even these political/human values.

CONCLUSIONS - NOW (1986)

The agreement reached by Canada's Environment Ministers in 1985 has included the real elements of cost minimization and political realities. The scientific analyses completed between 1982 and 1986 confirm the basic conclusions reached in the early decision:

(1) Removing S from the system translates over time into a decrease in both wet S deposition and flux between the U.S. and Canada;
(2) Our identification of the largest contributors to deposition in Canada has not changed;
(3) Cost optimization improves decision making when such decisions are made in finite steps; and
(4) Meteorological variability is very important when assessing change pointing out the necessity of baseline monitoring over an extended period.

REFERENCES

1. Altshuller, A.P. and G.A. McBean, Report of the United States - Canada Bilateral Research Consultation Group. Atmospheric Environment Service, 4905 Dufferin Street, Downsview, Ontario, M3H 5T4 (1979).
2. MOI Report of Work Group 2 on Atmospheric Sciences and Analysis, United States - Canada Memorandum of Intent on Transboundary Air Pollution (1982a).
3. MOI Report of Work Group 1 on Impact Assessment, United States - Canada Memorandum of Intent on Transboundary Air Pollution (1983).
4. Young, J.W.S. and R.W. Shaw, Atmospheric Environment 20, 189-199 (1986).
5. Shaw, R.W. and J.W.S. Young, Atmospheric Environment 17, 2221-2229 (1983).
6. Oppenheimer, M., Atmospheric Environment 17, 451-460 (1982).
7. Patterson, D.E., R.B Husar, W.E. Wilson and L.F. Smith, J. Appl. Met. 20, 70-86 (1981).
8. Young, J.W.S., Performance of LRTAP Models in 1982. Proceedings of the Second National Symposium on Acid Rain, Pittsburgh, Pennsylvania (October 1982).
9. MOI Report of Work Group 3B on Engineering Technology and Costs, United States - Canada Memorandum of Intent on Transboundary Air Pollution (1982b).
10. Barrie, L.A. and J.M. Hales, Tellus 36B, 333-335 (1984).
11. Shannon, J.D., Atmospheric Environment 15, 689-701 (1981).
12. Clark, T.L., R.L. Dennis, E.C. Voldner, M.P. Olson, S. Seilkop, and M. Alvo, The International Sulphur Deposition Model Evaluation, AMS 5th Joint Conference on Applications of Air Pollution Meteorology, pp. 57-60 (18-21 November 1986).
13. Shaw, R.W., Atmospheric Environment 20, 201-206 (1986).
14. Streets, D.G., D.A. Hanson and L.D. Carter, JAPCA 34, 1187-1197 (1984).
15. Batterman, S., M. Amann, H.P. Hettelingh, L. Hordijk and G. Kornai, Optimal SO_2 Abatement Policies in Europe: Some Examples. IIASA Working Paper, WE-86-42 (August 1986).

Source–Receptor Relationships and Control Strategy Formulation

David G. Streets
Argonne National Laboratory
Energy and Environmental Systems Division
9700 South Cass Avenue
Argonne, IL 60439

INTRODUCTION

The source-receptor relationship is an important outgrowth of our understanding of the physical and chemical processes occurring in the atmosphere. As we refine our knowledge of the mechanisms governing pollutant transport, transformation, and deposition, we gain a better understanding of the ultimate fate of different species in the complex web of physicochemical reactions in the atmosphere. When these reactions are simulated within the domains of time and space that are important for acidic deposition, and when prevailing climatic conditions are superimposed, a picture begins to emerge of the relationship between emissions at various source locations and deposition at various receptor sites.

Because of the great complexity of the atmosphere, our knowledge of the source-receptor relationship is currently limited. For individual events under specific meteorological conditions, we are beginning to gain a good mechanistic understanding of the processes involved; and for long time periods (seasonal to annual) and great distances (hundreds of kilometers), we are able to reproduce most of the features of measured wet sulfate deposition patterns using state-of-the-art linear models [1]. However, much more must be learned before these two capabilities are satisfactorily merged.

The importance of the source-receptor relationship to formulation of a control strategy is clear. With perfect knowledge of "who does what to whom," we would be able to identify precisely which sources should be controlled, and by how much they should be controlled, in order to achieve a desired level of deposition at a particular receptor site. In the absence of perfect knowledge, the technical questions take on policy overtones: do we know the source-receptor relationship well enough to take advantage of it in the design of a control strategy?

Figure 1 illustrates, in highly simplified form, the importance of the source-receptor relationship. In 1(A), meteorological conditions cause the receptor to be influenced primarily by dry deposition of species originating at the more-distant group of sources at the far left of the figure. In 1(B), meteorological conditions are different, and the receptor is affected primarily by wet deposition of species from the closer sources in the lower center. Clearly, if the conditions depicted in 1(A) were dominant, control of the sources at left would be necessary to reduce deposition at the receptor, whereas, control of the other group of sources would be required under the second set of conditions. Long-term climatic conditions would, of course, reflect a mix of the situations illustrated in 1(A) and 1(B); and it is the relative degree of influence of the two groups of sources on the receptor over a long time period that is embodied in the source-receptor relationship.

Published 1988 by Elsevier Science Publishing Company, Inc.
Acid Rain: The Relationship between Sources and Receptors
James C. White, Editor

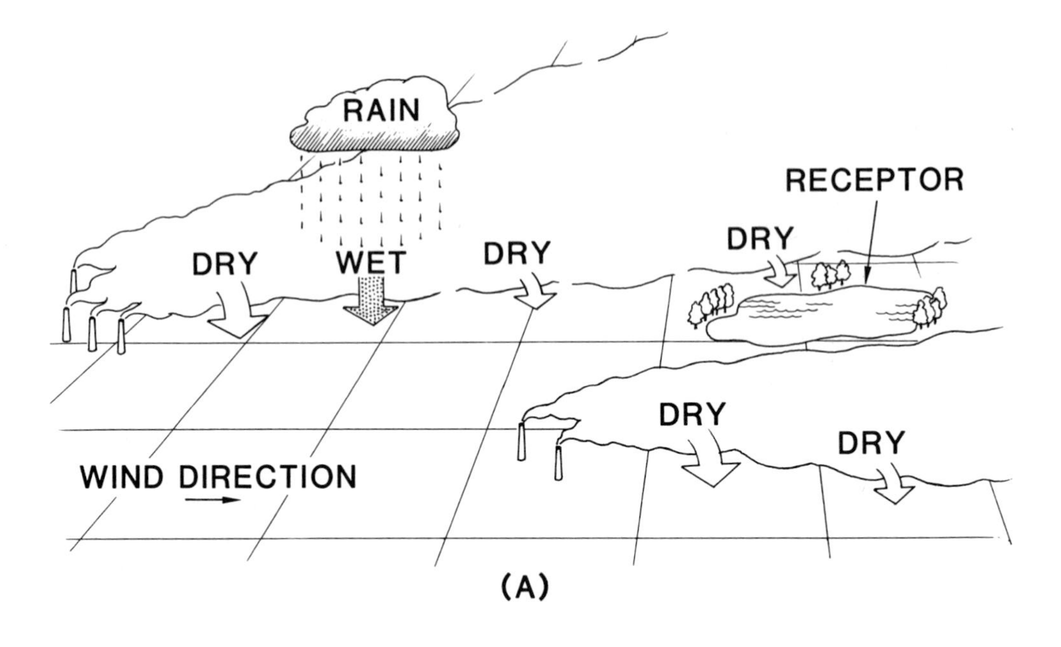

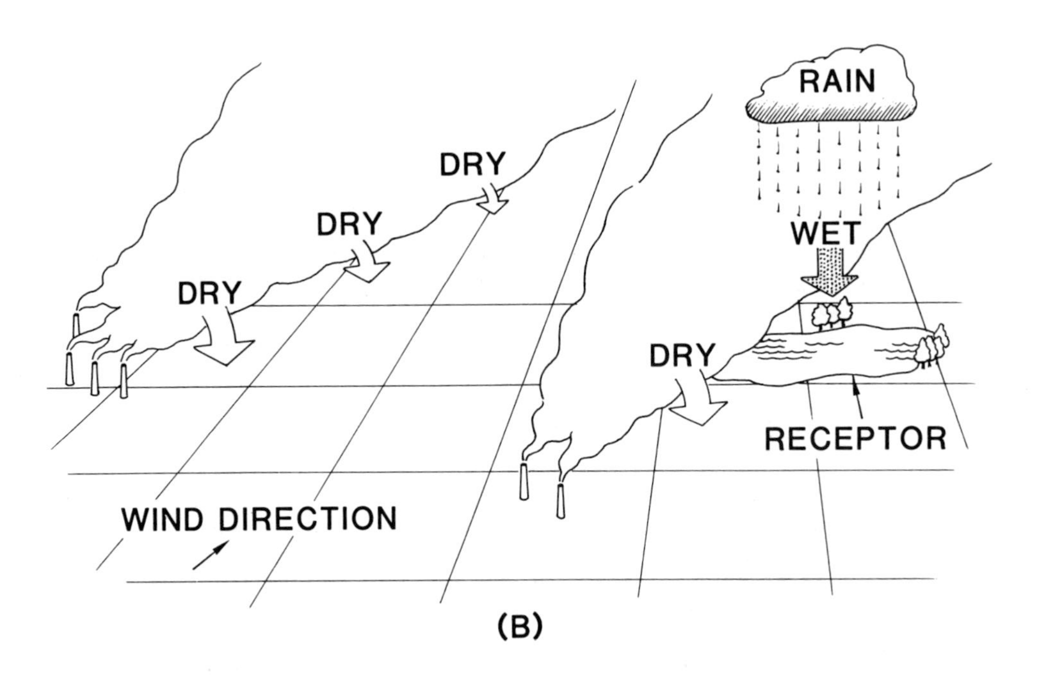

FIG. 1. Schematic illustration of varying meteorological influences on the source-receptor relationship.

Another important factor for control strategy decisions is the relative cost-effectiveness of reducing emissions at different sources. For example, returning to Fig. 1, it is possible that the relative costs or practicality of reducing emissions at the groups of sources at left and lower center could be greatly different. This should also be taken into account in developing a rational control policy.

For the purposes of the discussion that follows, we distinguish three types of strategy: emissions-optimized, in which one seeks to reduce emissions wherever they are greatest; cost-optimized, in which one seeks to reduce emissions wherever it is cheapest to do so; and deposition-optimized, in which one seeks to reduce emissions wherever it will lead to the most cost-effective reduction in deposition at a particular receptor site or set of sites [2]. This third strategy is the one that takes advantage of source-receptor relationships.

Almost all of the proposals introduced to Congress in the last five years have been of the emissions-optimized variety. It has been the thesis of several research groups that consideration of cost-optimized and deposition-optimized strategies could offer significant advantages over the "scattershot" emissions-optimized approaches. This paper focuses on work performed by Streets et al. at Argonne National Laboratory, with reference given to related work performed by Golomb and Fay at the MIT Energy Lab, by Shaw et al. at the Atmospheric Environment Service Canada, by McBean et al. at the University of Waterloo in Ontario, by Hordijk and coworkers at the International Institute for Applied Systems Analysis in Vienna, and others.

MODELING CAPABILITIES

As with all scientific investigations, control strategy analysis is limited by the tools that are available. All studies performed to date have used Lagrangian atmospheric transport and deposition models. These models can be described as semiempirical, in that they derive their credibility in part from describing observations rather than from providing detailed mechanistic explanations of molecular behavior [3]. More-advanced Eulerian models are under development, but their comprehensive, mechanistic treatment of atmospheric chemistry imposes severe computational limitations on control strategy analysis. Venkatram and Karamchandani state in their review of acidic deposition modeling that "(w)e should not discard semi-empirical models just because they are simple. ... We believe that complementary use of comprehensive and semi-empirical models will offer the greatest opportunity for understanding source-receptor relationships" [3].

The semiempirical Lagrangian models all assume linear transformation processes in the atmosphere, but we know that over certain temporal and spatial scales the transformations may be nonlinear. The implications of this assumption for control strategy formulation will be examined later in this paper.

To fully exploit the advantages of deposition-optimized strategies, we need to know more about ecosystem vulnerability than has been conclusively established by the research community. It is desirable to know which regions are most sensitive to acidic deposition, which chemical species are of most concern, and what limitations on deposition are necessary to protect the receptor. This third item involves elements of both science and value judgment; under such circumstances, it is prudent for the policy analyst to present the sensitivity of each type of control strategy considered to a range of environmental objectives.

Most studies addressing sensitive receptor regions have used the nine regions identified in the U.S.-Canada Memorandum of Intent (MOI) on Transboundary Air Pollution study [1,4]. These regions are shown in Fig. 2. Although presumed in the early 1980s to be sensitive to aquatic damage, these regions may no longer be appropriate in light of recent research findings on the sensitivities of surface waters in North America and the need for consideration of other receptor systems such as forests and materials. For control strategy purposes, there is a clear need for an up-to-date, multireceptor map identifying those regions particularly sensitive to acidic deposition.

With regard to the specification of adequate deposition rates for protection of sensitive receptors, estimates in the scientific literature range from wet deposition levels of 9 to as high as 30 kg/ha·yr of sulfate. Newcombe [5] reviewed the scientific literature pertaining to the levels of acidic deposition adequate to protect aquatic ecosystems. He found a seven-tiered hierarchy of harmful effects useful, ranging from "no effects observed" to "no natural reproduction of fish." Although 9 kg/ha·yr may represent the approximate lower limit at which biota have been lost from lakes, most control strategy studies have focused on the 20-kg/ha·yr value recommended in the MOI study [6].

Our knowledge of the costs to reduce SO_2 emissions below their current levels is much greater. Considerable study has gone into the capital and operating costs of flue-gas desulfurization equipment, the fuel premiums for low-sulfur coal, the costs of coal cleaning, and the relative costs of different fossil fuels and energy technologies. The work of ICF Inc. [7], the Congressional Budget Office [8], Carnegie-Mellon University [9], and Argonne National Laboratory [10] can all be consulted for appropriate cost information. Data for electric utilities are the most reliable, but information for other sectors is also available [11-14].

ANALYTICAL STUDIES

The key to efficiency in the design of control strategies lies in the fact that not all North American locations are equally sensitive to acidic deposition. Indeed, there is evidence to suggest that inputs of nitrogen and sulfur may, through a fertilizing effect, contribute to the productivity of certain field crops in the Midwest. The high alkalinity of soils in this region is sufficient to buffer the deposited acidity. On the other hand, other locations are known to be sensitive to acidic deposition because of low buffering capacity. In view of the very high cost of control measures, these factors argue for selecting sites for emission reduction to maximize the reduction in acidic deposition in sensitive regions while minimizing the cost of doing so.

1. Boundary Waters
2. Algoma
3. Muskoka
4. Quebec
5. S. Nova Scotia
6. Vermont/New Hampshire
7. Adirondacks
8. Pennsylvania
9. Smoky Mountains

FIG. 2. Locations of sensitive receptor regions in eastern North America. (Adapted from Refs. 1,4)

These ideas were first raised in the early 1980s. Goklany [15] asserted that "(m)ost of the SO_2 reductions obtained from ... untargeted controls over regions at least as large as the eastern U.S. will have little or no perceptible impact on reducing acid deposition in the Adirondacks and New England." He estimated that complete elimination of all SO_2 emissions in Ohio, Indiana, and Illinois would reduce deposition in New England (Region I) by only 6%, even though these three states contribute 30% of the SO_2 emissions in the eastern United States [15].

Trisko [16] recommended a "phased program of source controls, starting with nearby ... sources, and expanding later to more distant ... sources as necessary." Fay et al. [17] suggested that "considerably lower costs could be realized in an emission control scheme that is the more stringent the nearer the sources are to the environmentally sensitive areas." Argonne National Laboratory combined control cost models with atmospheric transport models, in an optimization framework, to analyze this possibility [13, 18-20].

The importance of the source-receptor relationship is immediately revealed by examining source-state contributions to "locally optimal" control strategies for specific sensitive receptor regions. By "locally optimal" control strategy we mean that combination of emission reductions that will reduce deposition at a specific single receptor site in the cheapest way possible.

Figure 3 shows state contributions to a locally optimal strategy for reducing wet sulfate deposition in the Adirondacks. Controls in this case

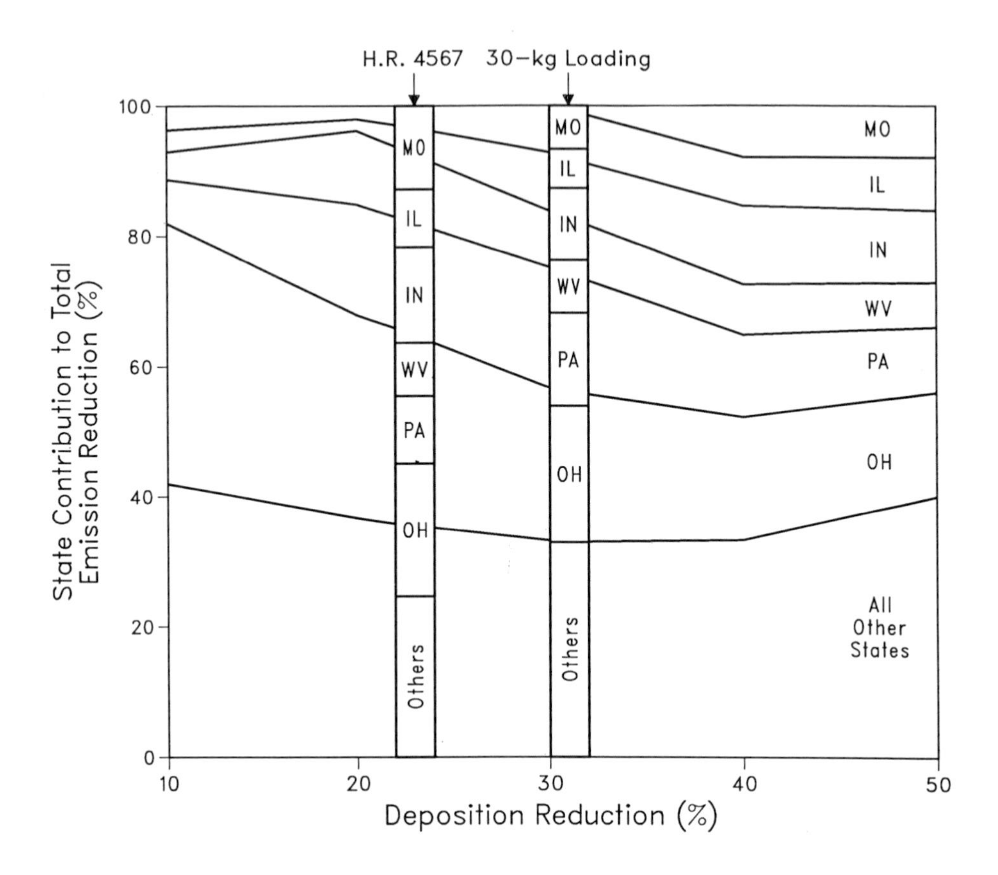

FIG. 3. State contributions to a locally optimal control strategy for reducing wet deposition of sulfate in the Adirondacks.

are limited to those available from coal-fired power plants in the United States. For low levels of deposition reduction (<25%), approximately 50% of the emission reduction would be drawn from just three states: Ohio, Pennsylvania, and West Virginia. These are large coal-burning states just upwind of the receptor region and that have relatively less expensive emission reductions available. As the desired level of deposition reduction increases (25-40%), contributions have to be raised from states increasingly further upwind: first Indiana, then Illinois, and finally Missouri. Ultimately, still more states have to be "levied" in the search for cost-effective contributions. At 50% deposition reduction (approximately the limit achievable through control of coal-fired power plants), 30% of the emission reduction is being drawn from the first three states, 30% from the second three, and 40% from all other states combined. For any level of deposition reduction, greater than 60% of the emissions reduction would come from just six states under a locally optimal control strategy.

Also shown on Fig. 3 are the state contributions for two specific control strategies, positioned according to the level of deposition reduction they would achieve under the same simulation conditions. These two strategies are (1) implementation of H.R. 4567, an acid-rain control bill introduced to the 99th Congress by Rep. Gerry Sikorski, and (2) a hypothetical strategy aimed at limiting wet sulfate deposition to below 30 kg/ha·yr in all nine sensitive regions [13,18]. The former reveals itself to be far from optimal for reducing deposition in the Adirondacks, "levying" too little emission reduction from Ohio and Pennsylvania and too much from the second tier of states. On the other hand, the 30-kg strategy matches remarkably well the ideal distribution of emission reductions.

Figure 4 shows the locally optimal strategy for the Smoky Mountains. Again, to achieve low levels of deposition reduction, neighboring states would contribute the greatest emission reductions: Georgia, Tennessee, Alabama, and Kentucky. These are states with many large coal-fired power plants, for which climatic conditions are conducive to pollutant transport to the Smoky Mountains. To achieve greater levels of deposition reduction, contributions from Ohio, Missouri, and more-distant states become necessary. Shown for illustrative purposes are the source-state contributions for H.R. 4567 and the 30-kg strategy. Neither is close to being optimal.

Inclusion of H.R. 4567 and the 30-kg strategy is not intended to show their unsuitability for these regions, because, of course, these strategies address many sensitive regions simultaneously. Rather, it is intended to show that these strategies are not optimal for some regions, and that a shift of focus from the source region to the receptor region may be potentially fruitful. The preceding analysis was conducted using the AIRCOST model of emission control costs and the ASTRAP model of atmospheric transport and transformation, within an optimization framework.

To demonstrate the potential economies offered by deposition-optimized (or "targeted") and cost-optimized strategies, relative to emissions-optimized strategies, Argonne analyzed several versions of Senate bills S. 768 and S. 769 from the 98th Congress [13,18,20]. The results are summarized in Table I. The Adirondacks region was chosen as the sensitive receptor region, and transfer matrices derived from six different long-range transport models were used to estimate deposition levels.

The two Senate bills are essentially emissions-optimized strategies to reduce annual SO_2 emissions in the United States by 9.0 and 14.9 million tons/yr, respectively, relative to 1980 levels. States with the highest emissions would be preferentially selected for the greatest share of the reduction requirement. Such measures are relatively expensive because of the economically inefficient allocation of emission reductions among the states. As Fig. 5 shows, control cost curves differ significantly among states. Thus, a cost-optimized version of the bills could reduce control costs by more than 50% while achieving the same level of emission reduction as would the bills. Cost-optimized strategies actually reduce sulfur deposition in the Adirondacks more than the bills would, because much of the cheapest emission reduction is to be found in the states in upwind proximity to the Adirondacks: Ohio, Indiana, Illinois, etc.

After determination of the deposition reductions that would be achieved by each bill, the six transfer matrices were used to calculate the most cost-effective way to achieve the same reductions in sulfur deposition in the Adirondacks, i.e., the deposition-optimized alternative.

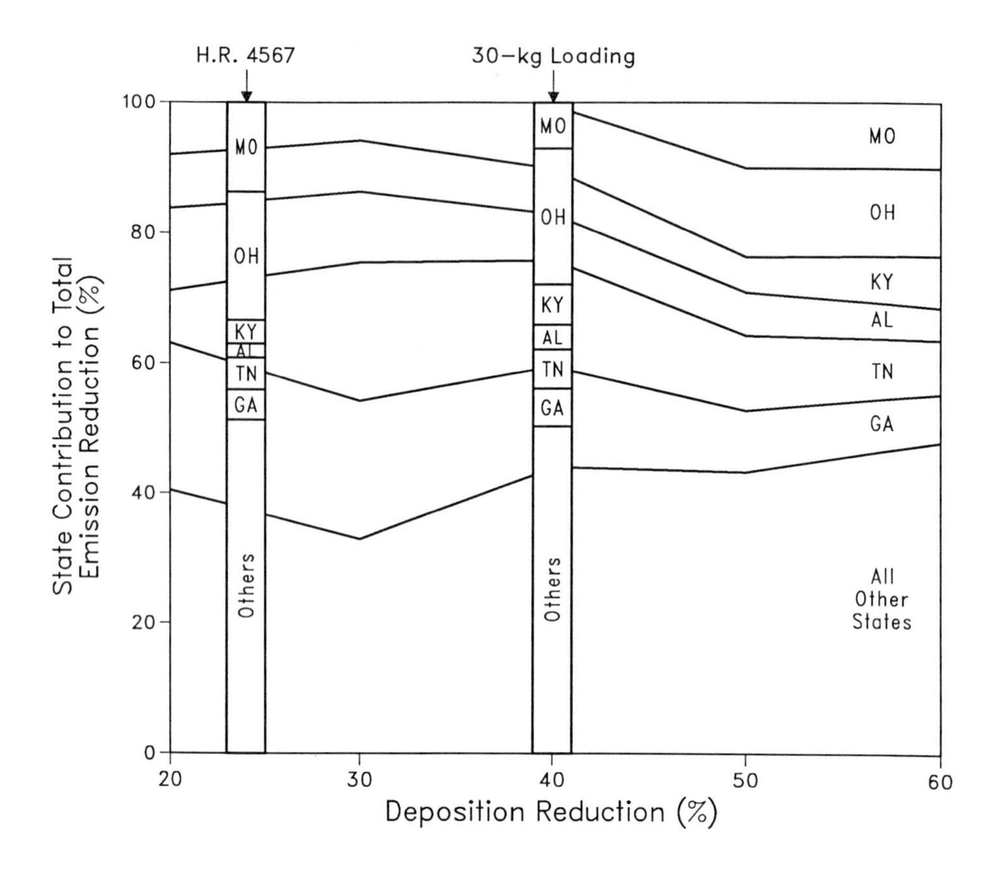

FIG. 4. State contributions to a locally optimal control strategy for reducing wet deposition of sulfate in the Smoky Mountains.

At an annual cost of $0.4-1.1 billion, the deposition-optimized strategy would reduce deposition in the Adirondacks by the same amount (21-26%) as S. 768, which costs $4.2 billion/yr. The ranges correspond to the results from the six different transfer matrices. Cost savings would therefore be 75-90%. The necessary reduction in SO_2 emissions would be 1.9-5.5 million tons/yr, compared with 9.0 million tons under the bill.

Similarly, at an annual cost of $1.3-3.2 billion, the deposition-optimized strategy would reduce deposition in the Adirondacks by the same amount (39-45%) as S. 769, which would cost $12.9 billion/yr. Again, the cost savings would be 75-90%. The emission reduction necessary under the deposition-optimized strategy would be 4.2-10.3 million tons/yr, compared with 14.9 million tons/yr under S. 769.

Generally, deposition-optimized strategies perform well in the other sensitive regions in the Northeast (Quebec, Nova Scotia, Vermont-New Hampshire, and Pennsylvania) when optimized for the Adirondacks. Figure 6 shows deposition reductions for each sensitive region under the three

TABLE I. Comparison of the effects of alternative control strategies.[a)]

Strategy	Annual Emission Control Cost ($\$10^9$)[b)]	Reduction in Annual SO_2 Emissions (10^6 tons)[c)]	Reduction in Total Sulfur Deposition in the Adirondacks[d)] (%)
Senate Bill S.768			
As proposed	4.2	9.0	23 (21-26)
Cost-optimized	1.9	9.0	26 (26-31)
Deposition-optimized	0.9 (0.4-1.1)	5.0 (1.9-5.5)	23 (21-26)
Senate Bill S.769			
As proposed	12.9	14.9	40 (38-46)
Cost-optimized	6.0	15.0	46 (45-57)
Deposition-optimized	3.0 (1.3-3.2)	9.8 (4.2-10.3)	40 (39-46)
20-kg Deposition Goal			
Including Muskoka	11.9	16.2	56
Excluding Muskoka	6.0	13.8	46
30-kg Deposition Goal	2.3	9.1	31

[a)]Numbers in parentheses represent the range of results given by six atmospheric transport models. The precision of deposition-change estimates is unknown, but the qualitative differences between strategies are believed to be reliable.

[b)]In real, levelized 1980 dollars.

[c)]In English short tons.

[d)]Modeled using the Advanced Statistical Trajectory Regional Air Pollution (ASTRAP) model, with 1980 meteorological data.

Source: Refs. 13,18.

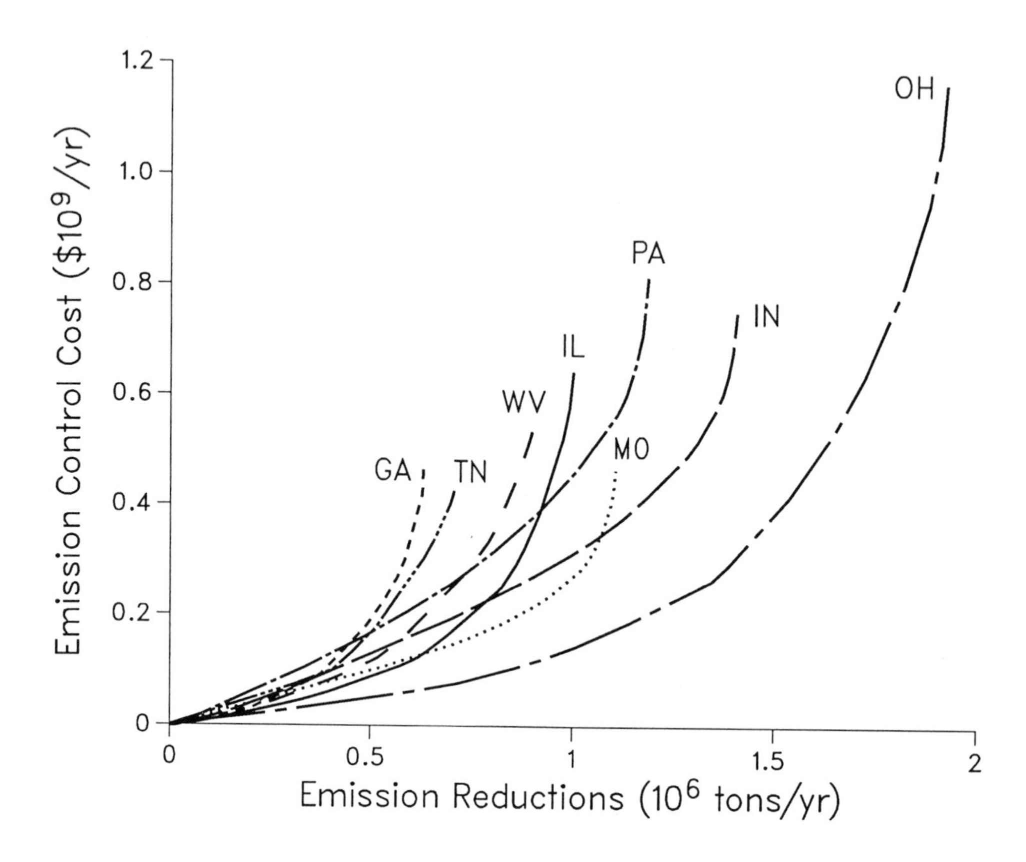

FIG. 5. Least-cost control curves for SO_2 emission reductions from coal-fired power plants in eight major states.

different strategies. Deposition is reduced less in the Boundary Waters region, Algoma, Muskoka, and the Smoky Mountains under the optimized strategy. Because deposition is currently low in the first two areas, this reduced protection may be of immediate concern only for the Smoky Mountains. The importance and urgency of this concern will depend on the sensitivity of ecosystems in the Smoky Mountains to acidic deposition. Muskoka, as is shown later, is a somewhat special case in this analysis because it is greatly influenced by Canadian sources. The policymaker would have the job of balancing the reduced protection afforded less-sensitive and insensitive regions against the substantial cost savings.

Figure 7 brings into sharp focus the advantages of the deposition-optimized strategy. It clearly illustrates the potential cost savings if the objective is to reduce sulfur deposition in the Adirondacks (and, with little loss of effectiveness, throughout the northeastern United States and

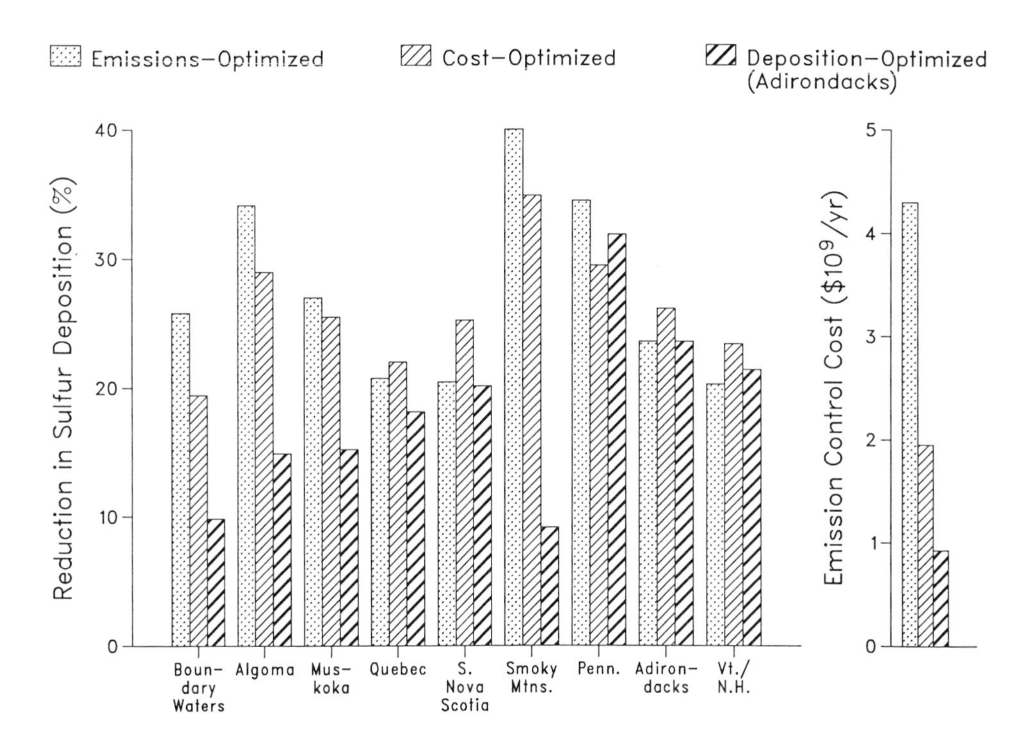

FIG. 6. Comparison of deposition changes in sensitive receptor regions under three alternative forms of control.

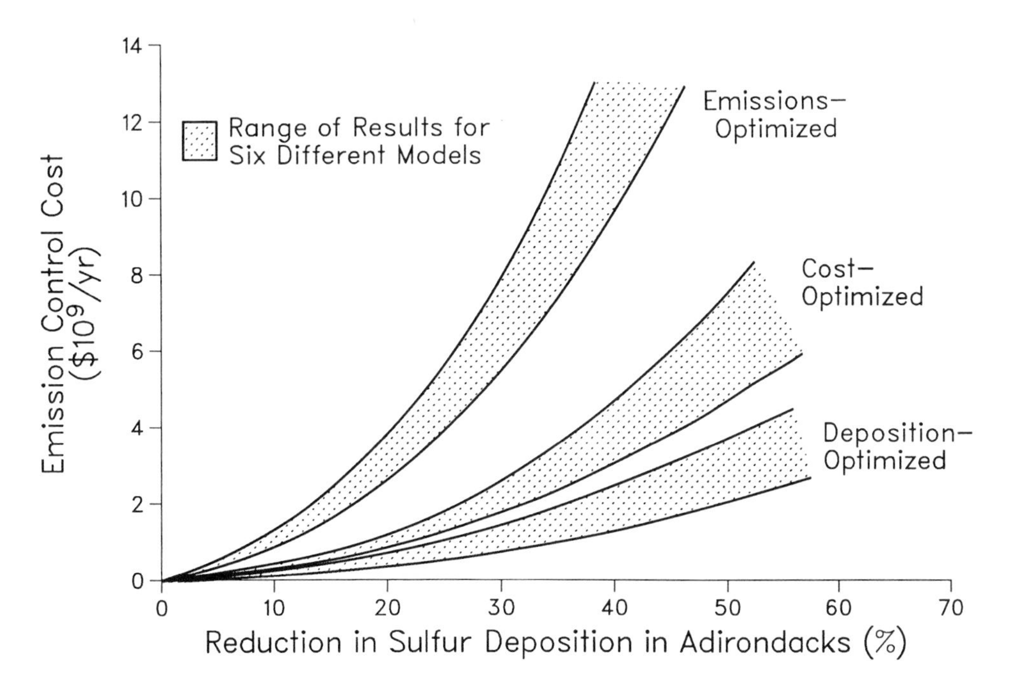

FIG. 7. Sensitivity of control strategy analysis results to choice of atmospheric transport model.

southeastern Canada). For example, the deposition-optimized strategy is clearly the least expensive way to reduce deposition in the Adirondacks by 40%. In contrast, the cost-optimized alternative would cost about 50% more and the emissions-optimized version would cost 350% more. Note also that the emissions-optimized approach is unable to achieve as much absolute reduction in deposition as the other strategies. Figure 7 affirms that these results are independent of the (Lagrangian) atmospheric transport and deposition model used.

Deposition-optimized strategies can also be designed to meet multiple objectives, such as achieving desired levels of deposition in many sensitive regions simultaneously at minimum cost. Analysis showed that wet deposition of sulfate could be reduced below 20-kg/ha·yr in all nine sensitive regions at slightly less than the cost of an unoptimized S. 769 (which would not achieve the 20-kg target in four regions) [13,18]. If the Muskoka region is excluded from this constraint -- because a large component of the deposition there is attributed to emissions from Canadian sources and this analysis did not consider controls on Canadian sources -- this goal could be achieved for half the cost of S. 769.

A wet-deposition goal of 30 kg/ha·yr of sulfate could be met in all nine sensitive regions for approximately 20% of the cost of S. 769. This strategy actually achieves levels much lower than 30 kg in the Adirondacks and the rest of the Northeast; Pennsylvania is the "binding" region (i.e., if the objective is met in Pennsylvania, it will also be met in all other regions).

Deposition-optimized strategies are somewhat sensitive to meteorological variability. Because these strategies are deliberately designed to take advantage of prevalent wind direction and precipitation patterns, variations from expected conditions may diminish their effectiveness. Figures 8 and 9 show the least-cost control curves that correspond to locally optimal strategies for the Adirondacks and the Smoky Mountains, respectively, evaluated on the basis of a six-year climate record [19]. Figure 8 shows that achievement of a 14-16 kg target in the Adirondacks, as recommended by Gorham et al. [21], could be very costly and conceivably impossible under certain meteorological conditions.

McBean et al. developed and used a screening model to assess deposition-optimized strategies for the eastern U.S. and Canada [22,23]. They simulated the installation of control technologies in a least-cost fashion on 235 large point sources representing approximately 67% of total man-made SO_2 emissions in the region. The Ontario Ministry of the Environment (OME) long-range transport model was used to develop transfer coefficients between source-receptor pairs. McBean et al. used 20 sensitive receptor regions, supplementing the nine MOI regions with 11 others in eastern Canada.

Using an LP optimization framework, McBean et al. evaluated least-cost strategies to achieve deposition targets of 20 and 25 kg/ha·yr of wet sulfate in every region [23]. They calculated that emission reductions of 3.5 million tons/yr (19%) and 10.8 million tons/yr (58%) would be necessary to achieve 25- and 20-kg targets, respectively. These values are lower than those estimated by Streets et al.: 9.1 million tons/yr (for a 30-kg target) and 16.2 million tons/yr (for a 20-kg target) [13,18].

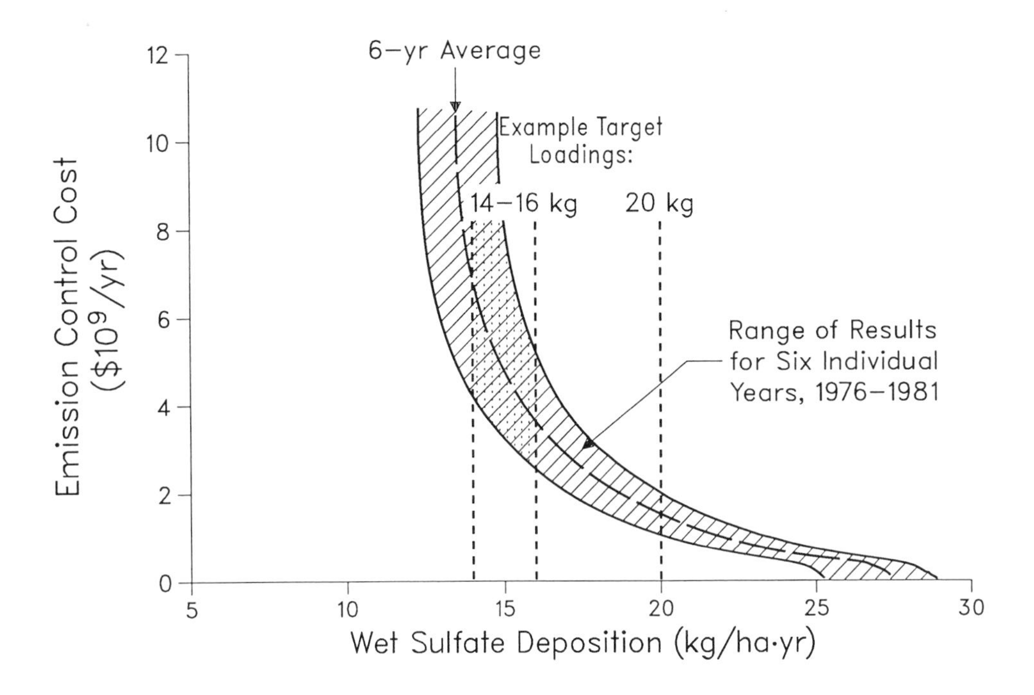

FIG. 8. Control cost curves for a locally optimal control strategy for the Adirondacks, evaluated for six years of climatic data.

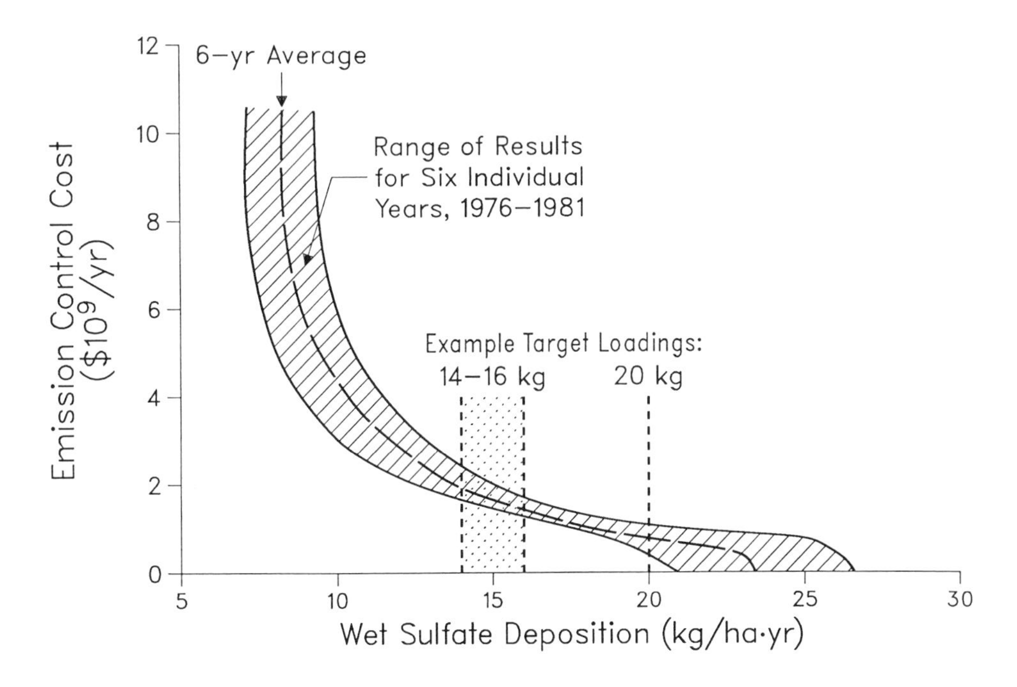

FIG. 9. Control cost curves for a locally optimal control strategy for the Smoky Mountains, evaluated for six years of climatic data.

These differences are explainable by the fact that McBean et al. included controls on Canadian sources, which facilitates achievement of the targets considerably at Canadian receptors. Streets et al. found the Muskoka region to be "binding" for a 20-kg target to such an extent that removing the constraint for Muskoka lowered the required reduction to 13.8 million tons/yr. The binding regions differ between the two studies and differ between targets within the same study, which reinforces McBean's statement that such differences are "... evidence of the complexity of the source-receptor linkage; emission abatement at one source to attain maximum allowed deposition at one receptor also has an impact on deposition rates at other receptors." McBean et al. further advanced the art of control strategy modeling by developing a probabilistic LP model that calculates the probabilities of exceeding deposition limits at specified receptor sites [24].

Shaw et al. [25,26] constructed a similar optimization model and used it to determine cost-effective ways to achieve a 20-kg target at many receptors. They concluded that "... most of the S emission reductions should take place in the Ohio River Valley, northern Appalachia, the lower Great Lakes region, and the provinces of Ontario and Quebec. It would be far less efficient, because of the weak meteorological link, to reduce S emissions elsewhere" [25]. This is consistent with previously cited work. Further, Shaw et al. affirmed that the selection of source regions (but not level of emission reduction) is relatively insensitive to year-to-year meteorological variability, uncertainty within a given model, and model-to-model differences.

Golomb and Fay examined more-efficient alternatives to the Mitchell bill, based solely on source-receptor considerations and not on control cost efficiency [17;27-29]. They calculated a "gain factor" for each strategy, defined as the change in deposition at a given receptor site divided by the change in regional emissions. The gain factors of the optimized strategies exceed that of the Mitchell bill by 25-90% [29].

Golomb and Fay also took advantage of source-receptor relationships in proposing seasonal and episodic control strategies. They reasoned that, because of the large differences in seasonal rates of wet sulfate deposition in eastern North America, reducing SO_2 emissions only in the summer half of the year would yield greater annual deposition reduction than reducing emissions by the same amount year-round. Similarly, because 10-15 rain events deposit about 60% of the annual wet sulfate, they suggested that reducing emissions in the dry periods before these rain events would improve control efficiency. While these statements are undoubtedly true, implementation of such strategies would be difficult from the standpoints of both technology and predictability. In addition, Knudson [30] was not optimistic about the magnitude of deposition reduction that could reasonably be achieved through episodic control of emissions.

The concept of using source-receptor relationships to improve control strategy efficiency has taken hold in Europe, with modeling activities underway at the University of York, the International Institute for Applied Systems Analysis in Vienna [31], the Research Centre for Water Resources Development in Budapest [32], and elsewhere. This level of activity suggests that many analysts are persuaded that we must pursue efficient approaches to acid rain control in order to minimize the economic burden required to achieve the desired environmental objectives.

This brings us back to the question of nonlinearity. Does any of the analytical work performed to date with linear atmospheric models have validity for control strategy decision making? Several atmospheric scientists have attempted to gain insight into the nonlinearity question without benefit of the comprehensive Eulerian models now under development. Samson [33] made an initial attempt to incorporate complex nonlinear chemical processes into the regional-scale Lagrangian model called ACID. His preliminary work showed that state contributions to wet sulfate deposition at Whiteface Mountain, New York, were not sensitive to hydrocarbon/NO_x ratios. Samson cautioned, however, that it was premature to make generalizations about the influence of nonlinear processes on source-receptor relationships for atmospheric sulfur species.

Exploratory work by Shaw and Young [34] predicted that a 50% reduction in sulfur emissions might yield a 35% reduction in sulfuric acid concentrations. Reducing NO_x emissions at the same time (by 50%) would reduce sulfuric acid concentrations by only 25-30%, i.e., greater nonlinearity; whereas the reduction of sulfur dioxide, nitrogen oxides, and hydrocarbon emissions each by 50% would reduce sulfuric acid concentrations by 40%. Except in the speciation between SO_2 and SO_4, there was little nonlinearity in total airborne sulfur.

Venkatram and Karamchandani [3] produced the first simulation of a regional emission reduction strategy using a nonlinear chemistry model called the Acid Deposition and Oxidant Model (ADOM). A simulated 50% reduction in SO_2 emissions produced a 50% reduction in wet sulfate deposition only far downwind of the source region (approximately 1000 km). Close to the source region, wet deposition was reduced by only 15-35%, and at moderate distances (hundreds of kilometers) by 35-45% -- "fairly linear" as the authors describe it.

CONCLUSIONS

Clearly, the results from the comprehensive Eulerian models are anxiously awaited. Efforts to date suggest that the results from Lagrangian models will have to be modified in light of nonlinear chemistry, but that these modifications are unlikely to alter the broad features of source-receptor relationships that we understand today.

Of the 50 or more bills introduced to Congress since 1981 for the control of acid rain, only one specifically recognized the importance of the source-receptor relationship in establishing a control program. Senator Durenberger's bill S. 2001 from the 98th Congress would have required the EPA Administrator to establish a priority list of sources to control that would "reflect scientific knowledge concerning the relationship of emission sources to receptor areas, for areas or natural resources particularly sensitive to acid deposition ...." [35]. It may be advantageous to reexamine such options if our goal is a cost-effective solution to the acid deposition problem.

ACKNOWLEDGMENT

This work was supported by the U.S. Department of Energy, Assistant Secretary for Environment, Safety, and Health, under Contract W-31-109-Eng-38. Because it has not been subjected to policy review, this paper does not necessarily reflect the views of Argonne National Laboratory or the U.S. Department of Energy.

REFERENCES

1. F.A. Schiermeier and P.K. Misra in: The Meteorology of Acid Deposition, P.J. Samson, ed. (Air Pollution Control Association, Pittsburgh 1984) pp. 330-345.
2. D.G. Streets, Environmental Progress, 5, 82-87 (1986).
3. A. Venkatram and P. Karamchandani, Environmental Science & Technology, 20, 1084-1091 (1986).
4. H.L. Ferguson and L. Machta, Atmospheric Sciences and Analysis Work Group 2 Final Report (United States-Canada Memorandum of Intent on Transboundary Air Pollution, Report 2F, 1982).
5. C.P. Newcombe, Environmental Management, 9, 277-288 (1985).
6. G.E. Bangay and C. Riordan, Impact Assessment Work Group 1 Final Report (United States-Canada Memorandum of Intent on Transboundary Air Pollution, Report 1, 1983).
7. ICF Incorporated, Analysis of 6 and 8 Million Ton and 30 Year/NSPS and 30 Year 1.2 lb Sulfur Dioxide Emission Reduction Cases (report prepared for the U.S. Environmental Protection Agency February 1986).
8. Congressional Budget Office, Curbing Acid Rain: Cost, Budget and Coal-Market Effects (U.S. Government Printing Office 1986).
9. M.B. Morrison and E.S. Rubin, Journal of the Air Pollution Control Association, 35, 1137-1148 (1985).
10. D.G. Streets, J.E. Vernet, and T.D. Veselka, Proposals for Acid-Rain Control from the 98th Congress (Argonne National Laboratory Report ANL/EES-TM-281, 1984).
11. M.E. Rivers and K.W. Riegel, Emissions, Costs and Engineering Assessment Work Group 3B Final Report (United States-Canada Memorandum of Intent on Transboundary Air Pollution, Report 3B Final, 1982).
12. Office of Technology Assessment, Acid Rain and Transported Air Pollutants: Implications for Public Policy (U.S. Government Printing Office 1984).
13. D.G. Streets et al., Controlling Acidic Deposition: Targeted Strategies for Reducing Sulfur Dioxide Emissions (Argonne National Laboratory Report ANL/EES-TM-282, 1984).
14. W.D. Baasel, J.G. Ball, and J.L. Meling, Chemical Engineering Progress, 47-53, June 1982.
15. I.M. Goklany, The Environmental Forum, 20-23 (November 1983).
16. E.M. Trisko, Public Utilities Fortnightly, 3-11 (February 3, 1983).
17. J.A. Fay, D. Golomb, and J. Gruhl, Controlling Acid Rain (MIT Energy Laboratory Report MIT-EL 83-004, 1983).
18. D.G. Streets, D.A. Hanson, and L.D. Carter, Journal of the Air Pollution Control Association, 34, 1187-1197 (1984).
19. D.G. Streets et al., Environmental Science and Technology, 19, 887-893 (1985).

20. D.G. Streets in: The Acid Rain Debate: Scientific, Economic, and Political Dimensions, E.J. Yanarella and R.H. Ihara, eds. (Westview Press, Boulder, Colo., 1985) pp. 173-218.
21. E. Gorham, F.B. Martin, and J.T. Litzau, Science, 225, 407-409 (1984).
22. M. Fortin and E.A. McBean, Atmospheric Environment, 17, 2331-2336 (1983).
23. E.A. McBean, J.H. Ellis, and M. Fortin, Journal of Environmental Management, 21, 287-299 (1985).
24. J.H. Ellis, E.A. McBean, and G.J. Farquhar, Atmospheric Environment, 19, 925-937 (1985).
25. J.W.S. Young and R.W. Shaw, Atmospheric Environment, 20, 189-199 (1986).
26. R.W. Shaw, Atmospheric Environment, 20, 201-206 (1986).
27. J.A. Fay and D. Golomb, Controlling Acid Rain: Policy Issues (MIT Energy Laboratory Report MIT-EL 83-012).
28. D. Golomb, J.A. Fay, and S. Kumar, Seasonal Control of Sulfate Deposition (Paper 85-1B.8, presented at the 78th Annual Meeting of the Air Pollution Control Association, Detroit, Michigan, June 1985).
29. D. Golomb, J.A. Fay, and S. Kumar, Journal of the Air Pollution Control Association, 36, 798-802 (1986).
30. D.A. Knudson, Evaluation of the Feasibility of a Regional Intermittent Control Strategy for Sulfur Dioxide (Argonne National Laboratory Report ANL/EES-TM-193, 1981).
31. S. Batterman et al., Optimal SO_2 Abatement Policies in Europe: Some Examples (IIASA Working Paper WP-86-42, Laxenburg, Austria, 1986).
32. J. Pinter, A Conceptual Framework for Regional Acidification Control, Systems Analysis and Model Simulation, in press, 1986.
33. P.J. Samson in: Trends in Electric Utility Research, C.W. Bullard and P.J. Womeldorff, eds. (Pergamon Press, New York) pp. 349-357.
34. R.W. Shaw and J.W.S. Young, Atmospheric Environment, 17, 2221-2229 (1983).
35. S.2001, A bill to amend the Clean Air Act to reduce interstate transport of pollutants, control acid deposition, and for other purposes, introduced by Sen. Durenberger to the 98th Congress, 1st Session, October 25, 1983, Part E, Sec. 183 (c)(3).

Questions and Answers following Control Strategy Session

Thursday, December 4, 1986

Q:

Do you make a conscious decision to base a lot of your analyses on political jurisdictions, such as states, for sources rather than looking at other definitions? I notice you use coal-fired plants, but scientifically, wouldn't there be a way of optimizing reductions by not focusing on political jurisdictions?

A: (Streets)

Yes, we do. We used the state boundaries primarily because those are the boundaries used in most legislation. Normally the required emissions reductions are allocated to states. So it's mostly on the grounds of political jurisdictions that we use that geographical split.

Q:

I would like to ask a political scientist's question. If there were no losers in this act, it would work. But I wasn't sure if I understood whether there are losers or not. By this I mean, if every state which has to reduce is going to have to reduce by less under your targeted procedure than they would have had to otherwise, then I think you have a chance of selling it. If some of them are going to have to reduce more than they would otherwise, then I think it's a serious problem.

A: (Streets)

I think the fundamental advantage of this kind of strategy is that we all, in aggregate, gain. In terms of particular state allocations, most states will reduce much less in their emissions. A few might reduce emissions more, because there are a lot of cheap emissions reductions to be had in some states, Ohio and a couple of others. And, of course, there are some losers in the receptor sense too. As I indicated, you don't get as much protection in certain areas that are far removed from the northeast, such as the boundary waters area or the far south. In terms of spreading the control costs, I personally believe that we should try to spread the costs over a number of states. In fact, I think we will pretty much have to do this with a targeted kind of concept, because it does put the burden inequitably on certain states. In order to take advantage of everybody having lower cost in aggregate, you would probably have to spread the costs over a number of states in a manner yet to be determined.

Q:

I would like to know on what you base the costs of reductions. Are you using a model such as EPA has used for ITF in the past? Can you point out if there is still a lot of interpreting to find out what the cost of low sulfur coal would be under major acid rain legislation? What is now thought to be cheap may not be as cheap as predicted.

A: (Streets)

I think that is a very good point. I think it is the major drawback of all models which have some coal price curves. If the price of low sulfur coal is effectively fixed, it is a drawback. As you say, the price would rise under a bill like this. But, on the other hand, there isn't that much difference between the low sulfur coal and the FGD option from any sources, so total costs might not be that much different; the choice could throw it either way.

Q:

Dr. Samson, I think you said on two occasions that the RADM model wouldn't be too good for source-receptor relations. Could you elaborate just a bit on that, particularly in view of the fact that other modelers have been running some simulations that certainly have some strong source-receptor implications?

A: (Samson)

I may have overstated what I meant. In terms of long-term source-receptor relationships, RADM is not designed to answer questions on the annual source-receptor relationships. It is very useful for looking at source-receptor relationships in individual storms, and on individual case studies we are getting very useful information. We may be able to infer something to the longer-term, but not for the overall long-term.

Scientific Studies Supporting Regulatory Development for Acid Deposition

George M. Hidy
Desert Research Institute
P.O. Box 60220
Reno, NV 89506

ABSTRACT

This paper contains a brief summary of presentations and discussions that took place at the CEI symposium on Source-Receptor Relationships in Acid Rain. The current state of knowledge was organized in terms of (a) needs for exposure characterization accounting for potential effects of acid deposition, (b) the formulation of source contributions to receptor exposure (source-receptor relationships -- SRRs) and (c) the regulatory implications of evolving knowledge of SRRs since 1980. The scientific results now available have not "hardened" information on possible ecosystem effects, particularly for terrestrial systems. Studies have given considerable detail about SRRs and their uncertainty. However, they do not appear to have changed the knowledge of broad average features of SRRs on a state or province scale, and larger. Given this situation, the development of scientific information on SRRs has not greatly improved the technical basis for decision making to regulate air pollution emissions for acid deposition control. Rule making for addressing this issue remains highly political in character. This is much the same situation that has existed since the early '80's.

INTRODUCTION

1986 ends another year of inaction dealing with the environmental issues of acid deposition. Since 1980, more than forty bills have been introduced in Congress for acid precursor emission control. None of these has reached the floors of Congress. The reasons for this inertia are complex. They relate to uncertainty in the significance of the effects of acid deposition, and to the high estimated costs of emission control. The potential effects have focused on surface water chemistry and aquatic biota changes, forest dieback, and material corrosion. Particular emphasis has been placed on the first two issues. Since 1980, continuing research has neither confirmed the severity and areal extent of change in surface waters in North America, nor the association of acid deposition with forest damage. In urban conditions, acidic pollutants have long been known to attack materials; in remote areas this effect remains problematical. The issue of public cost for deposition reduction has been a great concern, especially without compelling evidence of environmental effects. Indeed,

Published 1988 by Elsevier Science Publishing Company, Inc.
Acid Rain: The Relationship between Sources and Receptors
James C. White, Editor

perceived regional cost inequities have been a politically dominant factor inhibiting regulatory action. In the eastern United States where the issue is most visible to the public, the cost impact has been so large as to focus on targeted emission reduction as a means of cost control. Without evidence of widespread and severe environmental effects, the burden of new incremental control costs has received a cool reception relative to other priorities.

Possibilities for cost savings using targeted emission reduction have directed considerable effort toward establishment of relationships between deposition exposure at environmentally sensitive receptors, and source emissions (source-receptor relationships -- SRRs). These relationships are estimated from computer simulations (models), or from analysis of measurements. The reliability of the theoretical calculations was questioned in the early '80s, shifting interest to measurements. Analysis of measurements has its own limitations, partly from lack of quality field and emissions data, and partly from uncertain knowledge of atmospheric factors influencing SRRs required for data interpretation. This symposium has produced an updated survey of the capabilities to estimate SRRs and their limitations. This paper summarizes these reports in relation to needs for regulatory development.

EXPOSURE CHARACTERIZATION

Two speakers (Dr. Kulp and Dr. Likens, this volume) surveyed the information needs to characterize deposition exposure to terrestrial and aquatic ecosystems. Dr. Kulp reviewed the knowledge about crop and forest effects. Studies indicate that crops generally are not affected by acid deposition. However, research since 1980 has shown that montane forests in several alpine locations of eastern North America (ENA) are suffering dieback of a range in severity. Alpine forests that are suspected to be susceptible exist in the West, but exposures there are much lower than in ENA. Further, no evidence of effects have been reported except for localized oxidant damage. Scientists have hypothesized that air pollution, including acidic deposition, is potentially a factor in tree deterioration. This is especially the case for exposure in high altitude areas with thin, poorly acid buffered soils where trees are stressed by climate severities. Acid sulfate may be a factor in nutrient leaching from vegetation even if sulfur deposition does not exceed plant nutrient needs. The combination of leaf cracking induced by ozone (O_3) and moisture flow over leaves can accelerate leaching of nutrients. Although nitrate (NO_3^-) is an important nutrient deposited on forests, it also may be a factor in inhibition of tree development by stimulation of productivity at dormancy -- leading to adverse effects during freezing.

Particularly important to forest decline is believed to be the exposure to elevated oxidant concentration, especially O_3. Current work has focused on seven hour averages of ozone levels in relation to vegetation effects rather than the past emphasis on three hour maxima. Responding to the issue of forest decline, the National Acid Precipitation Assessment Program (NAPAP) has embraced O_3 as a key co-exposure factor for acid deposition. Studies of Dutch investigators also suggest that hydrogen peroxide (H_2O_2) may be a factor in damage. This is important because it complicates the strategy for control of chemicals potentially affecting forest growth. Instead of dealing with control of sulfur dioxide (SO_2) and nitrogen oxide (NO_x) emissions, for example, O_3 (and H_2O_2) reduction involve both NO_x and reactive hydrocarbon (HC) emission control. The atmospheric production of oxidant depends not only on the ambient concentration of these species but their ratio.

The potential for widespread effects of acid deposition on surface waters was the first to captivate the concern of the public through the Scandinavian research in the 1970's. Dr. Likens presented a viewpoint that would reinforce earlier perceptions of such damage in ENA. He noted that lake water chemistry is quite "complex" and seemingly ill understood. However, he also expressed concern for acidity (H^+) in a lake watershed, which acts to reduce alkalinity and to release toxic metals such as aluminum from soils or sediments into the water. In addition, he hypothesized that sulfate (SO_4^{2-}) in lakes and streams is of concern as an integral part of the acidification process, taking note that NO_3^- is biologically assimilated rapidly thus presenting no apparent problem. As a consequence, NO_3^- is considered an unimportant chemical in aquatic chemistry relative to H^+ and SO_4^{2-}.

In Likens' presentation, the old theme of the '80s repeated the warning that widespread acidification of lakes and streams can be expected in ENA under present exposure conditions. The author seemed to disregard a recent assessment in which the historical data do not support the perception that large numbers of lakes and streams have been affected by acid deposition [1]. Historical data, taken in balance, suggest that fewer than three of the documented historical samples of more than 100 lakes in the Adirondacks may have acidified from acid deposition between 1930 and 1980. Perceptions of change in other locations such as New Hampshire and Wisconsin are not borne out by historical data according to the NAS evaluation. Note, however, that a set of streams in the East, recently monitored, qualitatively reflect sulfate concentrations in step with reported emission changes in SO_2.

During the discussion, Likens noted that the deposition rate of SO_4^{2-} in ENA needs to be reduced to 10 kg/ha-yr to protect aquatic systems. This is substantially less than the 15-20 kg/ha-yr projected from scientific debate earlier [2]. Likens gave no scientific basis for this "new" threshold for SO_4^{2-} impact on the environment. However, a reduction in deposition to a new low level would involve even higher control costs than now projected.

Neither Kulp nor Likens gave adequate criteria for deposition measurement or estimates that relate to an exposure orientation. The investigators provided guidance for chemical species of concern, but they said little or nothing about temporal or spatial averaging required for effects assessment and specification of SRRs. No discussion of the significance of short term episodic exposures vs. long term cumulative effects was presented. Presumably, long term cumulative exposures still remain of principal concern for biosystem exposure in ENA.

ESTIMATING SRRS

After discussing the potential for ecological effects, the presentations gave a picture of scientific progress since 1980 in areas affecting improved prediction of SRRs. The audience generally was left with an impression of progressive work leading to exploration and definition of details, but little added substantiation or modification of the broad characteristics of SRRs deduced by methods used before 1983.

Given available information, seasonal and longer term average exposures from SO_4^{2-} and NO_3^- appear to be of primary concern. In ENA, variability of these averages is relatively small in magnitude and in (spatial) gradient. Shifts in total (wet and dry) deposition may take place with season; temporal changes are documented, at least indirectly, but spatial changes are only reported for cases of wet deposition. Even in a wet climate, precipitation occurs less than half the time. Thus, dry deposition has to be considered as a complementary factor to wet deposition. The closer to a source, the more likely dry deposition will be a large influence because of its dependence on ambient air concentration near the ground.

Dr. Hales (this volume) described in detail the complexities of chemical processes leading to atmospheric SO_4^{2-} and NO_3^- production. When unneutralized, these anions are strong acid producers in water. The importance of in-cloud processes was noted. Yet essentially no guidance was given for the influence of these chemical processes on changes to be expected in average deposition patterns over ENA, or their influence of SRR estimation.

The SRR calculations have relied heavily on air mass trajectory analysis. This method adopts linear rate processes, which are known to be an incorrect description of some important phenomena. The significance of the so-called linearity issue remains unresolved without field measurements for analysis of linear model performance. A concern for uncertainty and natural variability now pervades modeling studies, as discussed by Samson (this volume). Indeed, the broad features of SRRs by state, described by Streets and Young (this volume), are submerged in concern for the uncertainty in such projections. Even the more recent sophisticated schemes that use gridded, non-linear Eulerian models rather than the air mass (parcel) trajectory tracing are questioned. The model performance cannot be "proven" without a reference for comparison. Comparisons are generally made with specialized field data, which is largely unavailable.

In his presentation, Dr. Machta (this volume) stated that the analysis of historical measurements (primarily wet deposition) are qualitatively consistent with regional emission changes since the 1960's. However, observations of deposition summed from all sources offer no real way of segregating SRRs by target area, as yet. Isolation of SRRs by subregion appears to require some form of study involving tracer releases.

Release studies using chemically inert tracers have assisted in testing the inference of calculated air mass transport from models. But these results have not produced the link to calculated deposition rates. Establishment of an empirical relation between tracers and deposition patterns requires a very elaborate and expensive experimental verification [3].

Assuming that present modeling calculations yield reasonably reliable, broad SRR features, on the average, then little improvement in estimates may be projected except for short term events. Given this assumption, a range of SRR scenarios can be tested for their effectiveness. As Streets discussed, cost factors readily can be incorporated within the SRR calculation to generate a range of regulatory choices in which costs are optimized.

Although the ability to calculate SRRs has improved in efficiency and sophistication of theory, the capability for testing and verifying the calculations has not changed since 1980. The reason for this is that essentially no new, spatially and temporally comprehensive data have become available for confirming the calculated results. This is particularly the situation in which simultaneous, systematic measurements of wet and dry deposition are required. There is little prospect for new data acquisition and dissemination before the mid-1990s.

It is interesting to note that Dr. Kulp projected that the NAPAP results for assessment will not be available before the 1990s compared with the NAPAP legislative timetable of 1989. This amounts to an admission that characterization of effects, and the specifications of SRRs will require assimilation of studies well beyond the Congressional legislative limits. No one knows at this point whether or not Congress will agree to such delays, but they are to be expected given the complexity of scientific issues at stake.

REGULATORY IMPLICATIONS

The survey given in this symposium shows that there has been little progress in the broad aspects of knowledge about acid deposition that was not available in the early '80s. Many more scientific details are available now, and the uncertainties in interpretation of data are much better defined. Nevertheless, the information available for Congressional rule-making has improved little in its broad features over the intervening years. The situation is complicated, in part, by the lag in publication of assessment documents provided by NAPAP, and the lack of publication of a status summary in 1983. (The summary was drafted but never released to the public to accompany an encyclopedic survey of the issues by EPA in 1984 [4].) Release of NAPAP's awaited 1985 assessment report has not taken place as of year end 1986.

Congressional inaction has been reinforced by hiding behind the ambiguities of scientific information. Nevertheless, political action could be taken (at some risk) without strengthening the current scientific picture. Emissions reduction might be prudent for environmental protection; yet the public costs of uncertain protection apparently do not seem compelling, even today, when weighing the acid deposition potential for damage against many other environmental priorities, including hazardous waste disposal.

Scientific studies have been designed to reduce the risk of uncertainty in the decision making process. The SRR information acquired to date has not changed the conception of broad features in these relationships if statewide or EPA designated air quality sub-regions are examined in reference to key, large scale receptor areas. Research attempting to account for the hypothesized influence of atmospheric or surface processes does not appear to have the potential for markedly changing our knowledge of the broad features of SO_4^{2-} or NO_3^- deposition. The consequences of missing information to establish uncertainties in dry deposition or exposure by alpine cloud interception are not well understood because measurements are lacking. Such measurements will not be assimilated for some time. The level of detail needed in SRR estimation (and uncertainty) remains unclear. If more detail is deemed necessary, the more sophisticated mathematical models may be required for SRR assessment. Here a clear scientific policy or statement of specifications is needed. Cost analyses from the coarse, SRR based control options are available for a variety of scenarios. Unfortunately, their reliability cannot be determined at present without new field data acquisition and comparison.

Scientific evaluation of the issues awaits the early to mid-90's for completion. Given this framework, addressing acid deposition as an environmental issue is as political today as it was a few years ago.

There was discussion about the current legal framework of the Clean Air Act as it applies to acid deposition. The provisions of the Act can address the oxidant exposure issue as part of the review of the existing National Ambient Air Quality standard. Traditionally, EPA has included in its criteria documents for SO_2 and NO_x a review of exposure to dry deposition. It seems appropriate to extend that discussion to wet deposition. The crux of the standard setting process focusing on receptor conditions, however, cannot be addressed -- that is, the standard itself. At present, there is no means of quantifying a relation between ambient concentrations and deposition rates. Indeed a deposition rate "standard" or a comprehensive "threshold" for environmental damage remains beyond our state of knowledge.

The Clean Air Act currently deals with interstate and international air pollution issues. Yet, to date, the authority of the Act has unsuccessfully been applied to deal with such issues in relation to non-attainment of standards where one state's pollution may impinge on another. Except for O_3, the National Ambient Air Quality standards do not accommodate this possibility very well in that regional influences generally involve low ambient concentrations relative to local influences. Given these ambiguities, there is certainly motivation for additional, "clarifying" legislation. The direction taken towards an emission or discharge based approach characteristic of acid deposition bills is analogous to the approach taken in the Clean Water Act. This avoids the issue of a deposition standard at the receptor, facilitating the use of new control technology, but sidesteps commitment to a target for improvement. Indeed, the question of degree of improvement in exposure, per dollar spent, cannot be answered with any certainty.

A polarization of scientific opinion on acid deposition remains; calls for action still seem weakly justified by information on effects, compared with more pressing environmental issues. The dilemma remains with the politicians -- whose collective decision is difficult to predict in the coming years.

If the decision is made to proceed with emission controls, the public would be served best by legislation forcing new technology. Instead of resurrecting the bills that focus on retrofitting, a better approach would be to urge modernization through the use of new combustion technology. A particularly promising coal use technology employing gasification and combined cycle power plant operation is now available [5]. This scheme yields very low emission rates per ton of fuel in both SO_x and NO_x, and offers efficient incremental plant construction in 100-200 MW units that yield electric power comparable to current conventional technology. Application of new methods serves to direct industry into a new era of clean, efficient power generation. It also offers a badly needed means of systematically upgrading one of America's important smokestack industries to lead our progress in modernizing the Country's industrial capacity.

REFERENCES

1. National Academy of Sciences, Acid Deposition Long Term Trends (National Academy Press, Washington, DC 1986).
2. Swedish Ministry of Agriculture, Acidification Today & Tomorrow (Stockholm Conference on Acidification of the Environment, Stockholm, Sweden 1982).
3. Hidy, G.M., D.A. Hansen, and A. Bass, "Feasibility and Design of the Massive Aerometric Tracer Experiment" Report EA-4305 (Electric Power Research Institute, Palo Alto, CA 1985).
4. U.S. Environmental Protection Agency, The Acid Deposition Phenomenon and Its Effects. Critical Assessment Review Papers, Vols. I & II EPA-600/8-83-016 BF (U.S. Environmental Protection Agency, Washington, DC 1984).
5. Spencer, D.F., S.B. Albert, and H.H. Gilman, Science 232, 609 (1986).

What Does It All Mean?

Michael Oppenheimer
Environmental Defense Fund
257 Park Avenue South
New York, NY 10010

ABSTRACT

Scientific research over the last decade serves as a firm basis for developing acid deposition reduction policy. Understanding of source receptor relations is sufficient to support geographically broad-based reductions in precursor emissions. Improved source-receptor relations will facilitate the development of efficient reduction strategies, but major improvements from modeling cannot be expected for some time. Empirical methods, while promising, have received insufficient attention.

The object of this conference is to understand the importance of source-receptor relations, and to answer these questions about them: "What do we know? What do we need to know? And when will we know it?" I think we can answer these questions now based on what the previous speakers have said.

We have a firm general understanding of source-receptor relations; but we don't have much specific information. What do we need to know? My opinion is that we don't need to know any more to start controlling acid deposition, but it certainly would help us in the long term to know more, as underscored by the preceding presentations.

We could probably design a more efficient program if we did have a clearer understanding, so the research on this very interesting problem should proceed apace. How soon will our understanding improve? Here I am less sanguine. I think that a lot of progress can be made with the sort of empirical studies that Dr. Machta was talking about. On the other hand, based on what Perry Samson said and some of my own experience, I think we need to be a little skeptical that great progress is going to be made in the short term with theoretical long-range transport modeling. We've got to learn to live with current uncertainties for a long time, and make the best of it.

Let me just quickly go through some of the high points that support these views. Ralph Perhac indicated the importance of source-receptor relationships, and indicated that we would like to understand them even if emissions were simply cut across the board. I think that conclusion arises from his sense that, in the long term we may need to sharpen things and develop a directed program. Jake Hales gave a beautiful exposition of the complexities involved in the modeling. Basically he indicated that, with both wet and dry deposition, there is a large degree of mechanistic complexity and a large degree of uncertainty in the models, and that the measurements are not too reliable. It is extremely hard to justify a strong faith in transport models as long as data are so poor. In fact, I'd say that if there is a fault in the national assessment program, it is the lack of attention to the collection of empirical data, hand-in-hand with the modeling. I think model validation programs must be a lot more substantial than those which are currently planned.

Acid Rain: The Relationship between Sources and Receptors
James C. White, Editor

Larry Kulp spoke very strongly and interestingly about the possible role of ozone, but he said one thing I would quarrel with: "We wouldn't want to spend a lot of dollars on things that aren't important." He meant, I suppose, that we don't want to control the wrong pollutant. My view, based on what I heard from several of the speakers is that there are several pollutants doing a lot of damage, and that the damage is costly now. Therefore, it's going to be very hard for us to make a mistake by controlling the wrong pollutant whether we aim at sulfur, nitrogen, or ozone control. I think that's something we really needn't fear.

Gene Likens' presentation was interesting in several respects. He presented the links of a logical chain connecting emissions, deposition and the biological effects on aquatic systems. For watersheds, we have what approaches a quantitative understanding of how to counteract the acute effects of acid rain, at least on a short-term basis: reducing the deposition levels to something on the order of one-third of current levels of deposition. Likens noted that it would be nice to have additional information, but that such information is not really necessary to make decisions. He added, "We're getting a progressively more rational basis to make the decisions."

Perry Samson's was unusually skeptical in his attitude towards the models. I tend to agree with him that there are real limits to what we can currently expect from theoretical transport models.

Lester Machta gave an encouraging view on empirical approaches. There were a few speakers later that were very interesting, who basically said, "Look, forget the complexities, we're going to go ahead and take the attitude that we have to be pragmatic and we're going to do something." They agreed that, in spite of uncertainties, when you deal with the predictions on an annual average basis, and you look at substantial regions, models are relatively insensitive to the unknowns. This attitude allows us to make use of the source-receptor relationships. I want to emphasize here that some problems just don't get solved in the traditional sense, and the source-receptor problem is one of them. This problem will never be solved, but we will continue to make progress. We can be pragmatic in the meantime and try to move forward, using the knowledge that we have accumulated.

Reliable, hard and indisputable interpretations mutually acceptable to all interested in the source-receptor problem are going to be hard to come by, although I think empirical studies in particular will be useful in this regard. But our broad understanding of the relationship of regional sources of emissions to regional deposition is now certainly substantial enough so that we can start formulating policy.

We know that long-range transport occurs, and that local sources can be important under certain situations. We know that the atmosphere is grossly linear on a long-term and large-scale geographic average basis. We know that pollutants do travel farther than 600 miles, and we know that we may never understand point-to-point relation so let's not wait for them to be developed.

We know that the effects are broadly distributed, that we don't just have dead fish, that we have materials damage, damage to human health and damage to forests. Since these effects are widely distributed over broad geographical areas, we don't need to worry about point relationships. Damage is broadly distributed.

Small reductions targeted in specific areas based on the predictions of models probably will not have the consequences we expect. Large reductions targeted over a large number of sectors, broadly spread out geographically, with the models used for guidance, do have a chance of success.

My view is that we should go ahead and begin emissions reductions. Based on our understanding of aquatic systems, we need to make large-scale reductions. Our general knowledge of source-receptor relationships is being used right now, along with political and economic reasoning, to formulate reasonable reduction strategies. In the long run, source-receptor models will become progressively more important because the larger reductions, presumably needed at a later time, will be progressively more expensive, and it will be more important to have more refined information.

So my view is, let's get off the dime and start moving.

I would differ with George Hidy only in emphasis. I think you don't want to look at this problem only from the long view. There is a substantial body of evidence that, aside from damage to natural systems, there are quantifiable dollar damages to materials occurring now. Health damage, though less quantifiable, is occurring now. These are ongoing costs. Where quantified, they are in the same ballpark as the control costs. Controlling acid rain is not just a problem of hedging against further risks; it's a problem of stemming a hemorrhage of dollar damages being done today.

Questions and Answers following Conference Summary

Thursday, December 4, 1986

Q:

I feel a bit disappointed that this became almost an acid deposition forum, because the source-receptor relationship extends well beyond acidic deposition. I think Dr. Oppenheimer alluded to that very strongly when he mentioned other issues of environmental concern that have broad geographic limits. He said that because of these other concerns, we may not have to worry about an individual stack and an individual lake. If that is true, we may not be interested in the source-receptor relationship, but I can't agree with that and I don't think you would either. Dr. Likens introduced this broader aspect of the source-receptor relationship when he said that he would like to see a reduction to a level of around 10 kilograms per hectare per year. If society should opt for this level, it would turn to the use of the source-receptor relationship. We might then want to ask how much of what pollutants we would reduce and would need knowledge of the source-receptor relationship. My final plea is, if we are interested in the source-receptor relationship, it is not just an acid deposition issue. It covers a number of pollutants and a number of environmental issues, and if we are going to have an effective legislative regulatory strategy, the source-receptor relationship can and should play a role in it.

Q:

What would you think of a Congressional bill that put a lot of emphasis on oxidant reduction, in addition to a reduction in hydrocarbons, SO_x and NO_x? If Congress legislated on SO_x and NO_x without also including hydrocarbons, would it be unreasonable given the current state of the science?

A: (Oppenheimer)

We know that sulfur is driving the chronic acidification of surface waters, and we know roughly what we have to do to avoid that. That's the basis for underscoring sulfur, so far. But, without backing off on the need to do that sulfur reduction, it is certainly clear that oxidants such as ozone and hydrogen peroxide are responsible for a variety of ills in the atmosphere and on natural resources. In the long run, we are going to have to take a good hack at that problem as well. But when faced with a decision of which comes first, you fall back initially on sulfur by saying we have a better understanding of the quantitative benefits of its control. We have somewhat less of an understanding of the benefits of reducing oxidants as Larry Kulp pointed out. So I think the answer is, we have to do both of them, but given a preference we should deal with sulfur first.

A: (Hidy)

Your question is a very good one, but it is a very tricky one to answer. Within the current structure of legislation there are ways and means of implementing control of ozone and other oxidants. As a matter of fact, ozone control is being considered now. The concern I would have in trying to roll an acid rain bill is that it might get very complicated. The process of trying to determine the ambient standard for ozone in combination for those for SO_2 and NO_x might delay deposition models and lose everything in the standard development process. Science has recognized since the late 70s that ozone and oxidants play an important part in acid formation in the atmosphere and other important reactions. Treating these in a conventional way may not be appropriate with regard to legislation. So if I were king, I'd opt to keep them separate, set goals for deposition and set an ozone standard more on a combination of vegetation damage and health effects.

A: (Oppenheimer)

Let me add that, in the long term, we can make crude projections of where the levels of ozone are going in the next 50 years. As Larry Kulp mentioned, it has already roughly doubled above natural levels and I would project it is going to double again if we do nothing new regarding NO_x and hydrocarbons.

The Questions for Congress

Rosina M. Bierbaum, Robert M. Friedman,
and John H. Gibbons
Office of Technology Assessment, U.S. Congress
Washington, DC 20510

CONGRESS AND THE SOURCE-RECEPTOR RELATIONSHIP

We understand it is our charge to comment on how Congress has used information about "source-receptor relationships" in the past, and how they might in the future. OTA, as one of Congress's technical support agencies, has had the opportunity to observe Congress as it has handled the issue of acid rain for the past six years. Of course you must realize that there is no way we can pretend to represent the views of Congress. There are 535 Members, each with their own constituents and point of view. Thus, we will not try to guess what kind of questions Congressman Dingell, Senator Stafford or others would pose today. Rather, we will recount a rather brief history of acid rain control bills introduced into Congress over the last few years, particularly those aspects germane to the topic of this conference, "source-receptor" relationships.

Acid Precipitation Act of 1980

About the only law with a primary focus on acid deposition is the Acid Precipitation Act of 1980 (Title VII of the Energy Security Act of 1980, PL 96-294), signed June 30, 1980. Of relevance to this conference, the Act calls for an interagency effort that includes, among several others, programs for:

> "research in atmospheric physics and chemistry to facilitate understanding of the processes by which atmospheric emissions are transformed into acid precipitation;" and
>
> "development and application of atmospheric transport models to enable prediction of long-range transport of substances causing acid precipitation."

The results of the research are to be used to "formulate and present periodic recommendations to the Congress and the appropriate agencies about actions to be taken by these bodies to alleviate acid precipitation and its effects."

The First Control Proposals

Many members of Congress did not feel that the Acid Precipitation Act adequately addressed the concerns of their constituents. During October 1981, two bills were introduced to specifically reduce sulfur dioxide emissions in order to control acid deposition. The Mitchell and Moynihan bills (S. 1706 and 1709, respectively, of the 97th Congress) required about 35 percent to 45 percent reduction in Eastern U.S. SO_2 emissions. **Both approaches allocated emissions reductions directly to States, based on source characteristics alone.**

Published 1988 by Elsevier Science Publishing Company, Inc.
Acid Rain: The Relationship between Sources and Receptors
James C. White, Editor

Only a very general consideration of a "source-receptor relationship" appeared in the choice of control region. Emissions control was confined to the easternmost 31 States. The chosen 31-State region encompassed both the area of highest emissions (fig. 1) and highest deposition (fig. 2) -- plus a border of adjacent States, recognizing that pollution is transported across State lines. Moreover, at that time various studies had identified aquatic resources sensitive to acid deposition scattered throughout the region, from Maine to North Carolina.

Figure 1.—Sulfur Dioxide and Nitrogen Oxides Emissions—State Totals for 1980

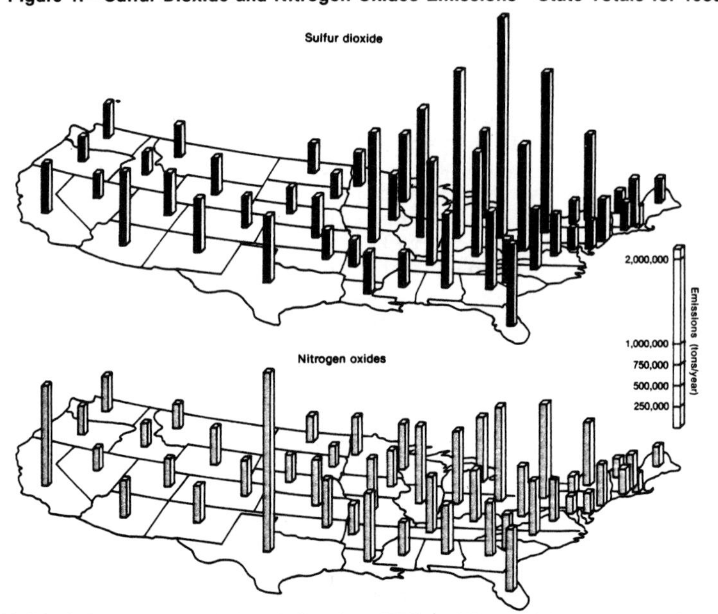

SOURCE: G. Gschwandtner, et al., "Historic Emissions of Sulfur and Nitrogen Oxides in the United States From 1900 to 1980," EPA-600/7-85-009 USEPA, 1985.

Discussions of the advantages and disadvantages of the two bills focused on: 1) the equity of the distribution of emissions reductions (a political judgment), and 2) the cost-effectiveness of emissions reductions (as analyzed by OTA and others using several available control-cost models). Atmospheric transport models were not yet one of the tools in the policy analyst's tool box. A third bill was introduced during the same week as the Mitchell and Moynihan bills: S. 1718 sponsored by Senator Dodd.

S. 1718, rather than directly specifying emissions reductions, attempted to clarify and strengthen existing provisions of the Clean Air Act that regulate interstate transport of certain air pollutants. The bill explicitly addressed the role for transport models when evaluating culpability.

Currently, EPA only considers interstate impacts that can be estimated by models approved under its modeling guidelines.

Because there are no Agency-approved long-range transport models, consideration of impacts is effectively limited to well within about 50 to 100 miles of the source.

Several States have objected to EPA's position and have attempted to compel EPA to use models and other techniques to assess long-range pollution transport. To date, the courts have deferred to EPA's position.

Figure 2.—Precipitation Activity—Annual Average pH for 1983

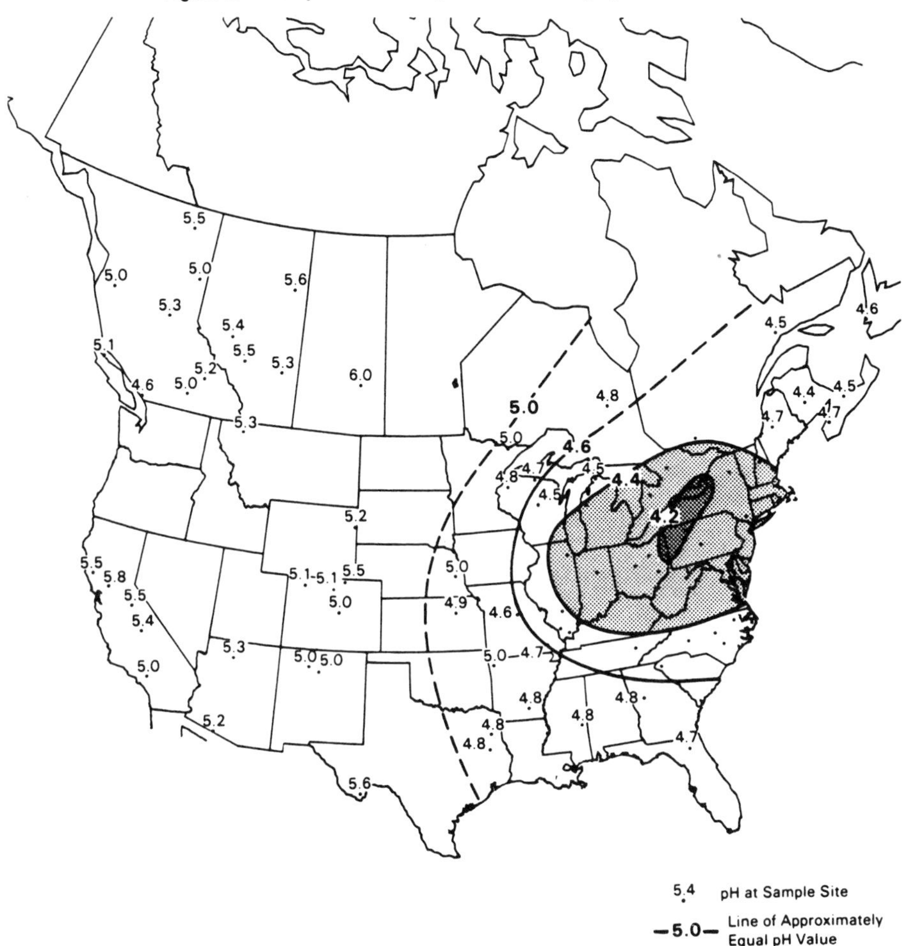

NOTE: Dashed lines are used when site density and length of data record are such that contour lines may not well represent the true patterns.
SOURCE: National Acid Precipitation Assessment Program, Annual Report to the President and Congress, 1984.

S. 1718 specifically addresses this controversy over the use of long-range transport models. It would prohibit EPA from relying **solely** on models when considering interstate pollution and transport. The bill would have allowed petitioners such as New York State to rely on the best scientific information available at the time -- wind patterns, monitoring data, and so on.

The bill cosponsors were stating that though they felt models could certainly provide some **useful information** about interstate pollution transport, lack of a definitive model was in no way to excuse the Administrator from issuing a decision.

EPA was to issue a decision on interstate pollution complaints using all available information, even though such information was incomplete and its accuracy uncertain.

Though most of the discussion of the Mitchell and Moynihan acid rain control bills concentrated on State-level allocation of emissions redutions and costs, the Senate Environment Committee did make one formal request for transport model analyses. During a hearing on the two bills, Senator Moynihan asked OTA to compare the deposition reductions that might be achieved from the two approaches.

Though many were shocked at that time, we believe that the Senator's request was reasonable. The Committee had been presented with analyses of differences in costs and emissions reductions mandated by the bills. Eventually they would have to accept one or the other (or neither) of the two proposals, but not both. Could source-receptor models help Congress decide between the two bills? As it turned out: Not very much. The bills were so close that models could provide little additional insight -- of itself a useful piece of information. Thus the debate remained focused on equity and cost-effectiveness concerns.

Though about a dozen emissions control proposals were introduced over the two years following the Mitchell, Moynihan, and Dodd bills, to our knowledge none made detailed use of the source-receptor relationship. The concept resurfaced in a bill introduced by Senator Durenburger in October 1983. S. 2001 would have required the EPA Administrator to publish a "priority list" of emissions reductions. Two of the specified criteria implicitly require the use of models.

The bill states that the priority list shall: 1) "assign highest priority to ... sources which will achieve the greatest degree of emissions reduction for each dollar" of public cost (as specified by a formula in the bill), and 2) "reflect scientific knowledge concerning the relationship of emission sources to receptor areas for areas or natural resources particularly sensitive to acid deposition."

Similar to most of the other control proposals, the bill specifies the total required emissions reductions (10 million tons of sulfur dioxide). But unlike other proposals, the allocation of emissions cutbacks is determined by EPA following a series of technical analyses, including consideration of source-receptor relationships.

As attractive as such an approach may sound to technical folks, the bill attracted few Congressional cosponsors. There may be a lesson here.

The Durenburger proposal recognizes that there is a role for source-recptor models in acid rain control. The bill assigns models and modelers to the role of advisors **to the EPA Administrator.** The Administrator is to weigh this information along with many other considerations and construct an overall priority list for emissions reductions. Though many might have supported the role of models and modelers as advisors (not as decisionmakers), few advocated leaving the eventual allocation of emissions to the EPA Administrator.

The 99th Congress

During the 99th Congress, attention focused on three acid rain control proposals: H.R. 4567 on the House side and S. 2203 and S. 2813 in the Senate.

Again, about the only consideration of a "source-receptor relationship" appears in the choices of control region. Both H.R. 4567 and S. 2203 propose national emissions controls. S. 2813, similar to the original Mitchell Bill from 1981, confines most of the stationary source controls to the easternmost 31 States. All of these approaches allocate emissions reductions to sources or States based on emission rates, rather than detailed knowledge of source-receptor relationships.

What might one conclude from all this? While there **have** been a few attempts over the past five years or so to include detailed source-receptor relationships in acid rain control proposals, Congress most certainly has **not** been attracted to this approach.

CHARACTERISTICS OF THE POLICY PROBLEM

Though over five years have passed since the introduction of the first acid rain control bill, the highly controversial public policy issue of whether or not to take action to further control pollutant emissions still remains. Several characteristics of transported air pollutants have fueled the public policy debate. **First**, scientific uncertainty exists both over the extent of the consequences of transported air pollutants, and the effectiveness of pollution control. **Second**, significant disagreements in values exist over how to balance the costs of controlling pollutants with the environmental costs and risks posed by pollution. **Third**, and of great political importance, transported air pollutants pose a distributional problem, having intersectoral, interregional, international, and intergenerational equity aspects.

Disagreements over Facts

Debate over scientific understanding of transported air pollutants is perhaps the most visible aspect of the policy controversy. Scientific uncertainties affect several policy concerns, including: 1) making air pollution control policy as fair as possible, i.e. providing some legal recourse to those bearing the risks of damage, and (if a control program is adopted) distributing costs of control fairly; 2) the inter-generational implications of the risks of cumulative and irreversible damage; 3) weighing the risk of damage against gains that might be achieved by waiting for better information or improved technology; and 4) assuring that the benefits, in the broadest sense, of a control program justify the cost.

Disagreements over Values

Even if a scientific consensus existed on the magnitude of the problem of transported pollutants, policy choices would still be complicated by lack of agreement over how to maintain economic development while protecting the environment. Although the concept of a tradeoff between these two values is widely accepted, various individuals and groups differ sharply on where the balance should be struck.

Distributional Issues

Additional value conflicts arise from the distributional aspects of transported air pollutants, further complicating the policy dilemma. Winds carry pollutants over long distances, so that activities in one region of the nation may contribute to resource damage in other regions. Many of these activities primarily benefit the source region, while some of their costs, in terms of resource damage caused by the eventual disposal of their waste products, are incurred by other regions. **Long-range** pollution transport thus redistributes benefits and costs across regions. Programs to control transported air pollutants will also have interregional distribution aspects.

It is because of these distributional aspects that an understanding of the relationships between sources and receptors can offer much to the policy debate. At the same time, however, we must recognize that several types of distributional issues -- not merely source-receptor relationships -- must be considered.

LEGISLATING EMISSIONS REDUCTIONS: DESIGNING A CONTROL PROPOSAL

If Congress in its collective wisdom decides the risk of ecological damage from transported air pollutants outweighs the costs of control, eight questions must be answered to design a control strategy. The OTA report painstakingly outlines many options and answers under each question. Rather than list all possible answers, we will describe how the major bills under consideration today respond to these questions.

Question 1: Which Pollutants Should Be Further Controlled?

All of the major bills go after sulfur reductions first. In the eastern U.S., sulfur compounds are responsible for about two times more acidity in precipitation than nitrogen compounds. Sulfur compounds are known to be related to lake and stream acidification, visibility degradation, materials damage and are implicated in human health effects. We have good emission inventories for sulfur (because it is measured in fuel delivered to a plant); therefore emission reductions are verifiable. All of the major bills of the 99th Congress, recognizing that nitrogen oxides and more recently ozone, may also be problems, seek to achieve some reductions in NO_x; some seek reductions in hydrocarbons as well. (As an aside, insights from transport modellers and atmospheric chemists will be sorely needed to design a workable regional ozone control strategy, if Congress decides such control is desirable.)

Question 2: How Widespread Should a Control Program Be?

Three regions are potential candidates (see fig. 2): 1) the bullseye of approximately 21 States receiving the greatest levels of acidity (i.e., precipitation acidity lower than pH 4.5); 2) a 31-State region (all States east of and bordering on the Mississippi River) incorporating a band of States around the region of greatest deposition; and 3) the continental 48 States. All of the bills in the last two

Congresses identify either a 31-state region or the continental United States, implicitly defining acid deposition as a large-regional problem requiring a regional solution.

Question 3: What Level of Pollution Control Should Be Required?

A decision on how much to reduce emissions involves two important components: (1) the scientific question of the relationship between emissions reductions and resource protection and (2) the policy question of the socially desirable level of protection. Thus, the types of questions that are asked include: what is an acceptable level of resource exposure?, how much do you have to reduce emissions to get there?, and how much do you feel you can afford to pay? Most bills to date have proposed emissions reductions in the range of 8 million to 12 million tons of SO_2, some requiring an additional 4 million tons of NO_x. Most of these proposals would cost in the range of \$3 billion to \$6 billion or more per year and presumably reflect the conviction that avoiding resource damage is worth at least that amount. The 35 to 50 percent reduction in sulfur dioxide would come close to protecting most aquatic resources in the Eastern US.

Question 4: What Approach to Control Should Be Adopted?

Options include: **source-oriented** (directly specify emissions reductions or specify use of specific control technologies); and **receptor-oriented** (establish an environmental quality standard). A sophisticated understanding of source-receptor relationships is required for the latter approach -- probably more sophisticated than we can claim at present. Thus, almost all bills proposed to date have directly specified emission reductions.

Question 5: By What Time Should Reductions Be Required?

Some have proposed reductions by the early 1990's. Others have proposed a two-phased approach to allow the results of the ongoing research program to be incorporated into the control plan. Realistically, a Congressional decision to require substantial further controls is likely to take 6 to 10 or more years to fully implement from the day of decision. Smaller levels of emissions reductions that require only minimal capital investment might be achieved somewhat more quickly.

Question 6: How Should Emissions Reductions Be Allocated?

Directly to sources or to States? By Congress or EPA? Different bills use all possible permutations of these choices. One could allocate emission reductions using a model and a detailed source-receptor relationship, but almost all the bills have instead specified emission reductions to States or directly to sources **based on emission rates** (expressed in pounds of pollutant per million Btu of fuel burned).

Question 7: Who Will Pay the Costs of Emissions Reductions?

There is no consensus on this answer. Allocating requirements for emissions reductions and allocating the **cost** of those reductions are two distinct issues. Both the traditional "polluter pays" principle and efforts to distribute costs through some taxing mechanism have been proposed in the various bills.

Some bills stop at this point; others seek to mitigate the undesirable employment and economic effects of a control policy.

Question 8: What Can be Done to Mitigate Undesirable Effects of a Control Policy?

Some bills mandate technology-based control measures (scrubbers) in whole or part to protect the high-sulfur coal market. A general tax on electricity to spread the cost to a larger group than those actually involved in reducing emissions has also been proposed. It should be noted that both of these last two questions -- which are probably the most contentious issues at this point in the debate -- require **no** source-receptor information.

Choosing from the Menu

Congress can specify a control program by considering each of the eight "decision areas" we just summarized. Obviously, many alternatives are possible. There is a continuum of available choices ranging from modest, interim reductions to large-scale control programs. Each strikes a different balance between the risks of future resource damage and the risks of an inefficient program.

The process of choosing between risks and benefits associated with each of these decisions is not only a regional one but also a national one. The lady at the lunch counter in figure 3 reflects not only the choices that each Member must make, but also the Nation as a whole.

A variety of proposals has been introduced to address acid deposition in the 97th, 98th and 99th Congresses. Some of the early bills called for a multi-faceted approach to transported pollutant problems such as accelerating the federal research effort and funding mitigation of the effects of acid deposition on acid-altered bodies of water as well as some emissions reductions.

Interestingly, almost all bills introduced in the last Congress converged on answers to the first 4 questions: large emission reductions of primarily sulfur dioxide over large regions of the country using a source-oriented allocation formula. Most of the bills proposing direct action would have established a 31-State or 48-State control region, and required reductions in annual sulfur dioxide emissions ranging from 8 million to 12 million tons below actual 1980 emissions levels by some point in the early 1990's.

The remaining questions -- those with an economic and equity component -- have been answered in many different ways in the various proposals. Moreover, we see no signs of convergence. While about 200 Members of Congress cosponsored an acid rain bill this year, over 300 members of Congress did not... and the situation remains contentious.

Figure 3

Drawing by S. Harris; © 1979
The New Yorker Magazine, Inc.

SOME CLOSING THOUGHTS

We would like to close this discussion with some more general thoughts about why air pollution has become a major policy challenge.

1. One reason the issue of governance and air quality is so traumatic to us as a Nation is that we must admit to ourselves the demise of the notions of "free goods." Such a deeply rooted historical feeling is not easily compromised.

2. Controlling air pollution is becoming **both** technically and jurisdictionally more complicated. Things get rapidly more complex as we progress from large point-source problems (e.g., total particulates), to dispersed but still local problems (e.g., carbon monoxide), to regionally transported and transformed pollutants (single, multiple, and complex combinations), to climate modifiers with global implications. It's a far cry to go from local action to global action!

3. Many important technical issues and uncertainties are

present, some of which reflect ignorance of the underlying science and some the changing nature of technology over time. Both kinds of uncertainty make policy choices more contentious. For example, policymakers must decide whether it's worth the risk of cumulative damage to defer control investments in order to take advantage of advancing technology.

4. Even more difficult, though, policymakers also must address some **normative** issues including:

   o How to choose amongst competing needs in order to invest resources more productively (even **within** the domain of air pollution).

   o How to decide what is an "equitable" distribution of costs and benefits between the economy and environment as well as among regions, specific economic sectors, and generations.

5. Finally, policymakers must also take into account several **international** concerns, including:

   o The general problem of economic competitiveness. Do U.S. environmental regulations give foreign competitors an unfair advantage?

   o U.S. border problems concerning air pollutions remain important issues for discussion and resolution.

Speakers and Participants

Beth K. Baird
Colorado Department of Health
Denver, Colorado

Morton L. Barad
Barad Consultants
Belmont, Massachusetts

W. Richard Barchet
Battelle-Pacific, NW
Richland, Washington

Laurie Batchelor
Alabama Power Company
Birmingham, Alabama

Dr. Gordon A. Beals
Consolidated Edison
New York, New York

J. Christopher Bernabo
Science and Policy Associates
Washington, D.C.

Neeloo Bhatti
Yale University
New Haven, Connecticut

Blair Blankinship
University of Maryland
College Park, Maryland

Michael L. Bowman
Maryland Power Plant Research
Annapolis, Maryland

Gary Breece
Georgia Power Company
Atlanta, Georgia

Helen Briassoulis
University of Cincinnati
Cincinnati, Ohio

Richard Burkhart
US Environmental Protection Agency
Boston, Massachusetts

Jeff Burnam
Office of Senator R. Lugar
Washington, D.C.

Jeff Caldwell
Johns Hopkins University
Baltimore, Maryland

Charles Carter
US Environmental Protection Agency
Washington, D.C.

Dean Carpenter
Oak Ridge National Laboratory
Oak Ridge, Tennesee

Norman Champine
OMYA, Inc.
Proctor, Vermont

Carol A. Charnigo
Syracuse University
Syracuse, New York

Eric Chasanoff
National Broadcasting Company
New York, New York

Sheldon Cheney
National Agriculture Library
Beltsville, Maryland

Carol Cole
Coal Week Magazine
Washington, D.C.

Timothy Crawford
Department of Commerce
Idaho Falls, Idaho

Nolan A. Curry
American Institute of Chemical Engineers
Troy, New York

Edward P. Curtis, Jr.
Center for Environmental Information
Rochester, New York

Nancy S. Dailey
Oak Ridge National Laboratory
Oak Ridge, Tennesee

Richard R. D'Auteuil
Buckeye Power Company
Columbus, Ohio

Mark C. Darrell
American Gas Association
Arlington, Virginia

Carl C. Dirkes
Steep Rock Resources
North York, Ontario Canada

Wilson P. Dizard III
Coal Outlook
Arlington, Virginia

Mary Beth Donnelly
Newmont Mining Corporation
Washington, D.C.

L. Clint Duncan
Central Washington University
Ellensburg, Washington

Dr. Ray W. Effer
Ontario Hydroelectric
Toronto, Ontario Canada

Mohamed T. El-Ashry
World Resources Institute
Washington, D.C.

Mark T. Ellis
Utah Bureau of Air Quality
Salt Lake City, Utah

Giles Endicott
Ministry of Environment
Toronto, Ontario Canada

James A. Fay
Massachusetts Institute of Technology
Cambridge, Massachusetts

Howard J. Feldman
Middle South Services
New Orleans, Louisiana

Elizabeth Field
Department of Environmental Regulation
Tallahassee, Florida

James Flanagan
Environmental Monitoring
Chapel Hill, North Carolina

Richard H. Forbes
Eastman Kodak
Rochester, New York

Monique Fridell
Renault, Inc.
Washington, D.C.

Shirley Fox
Department of Natural Resources
Talahassee, Florida

Lisa Frost
National Wildlife Federation
Washington, D.C.

N. E. Gallopoulos
General Motors
Warren, Michigan

Josie Gaskey
Allegheny Power Service
Greensburg, Pennsylvania

Norman E. Gauch
American Fishing Tackle Manufacturers
Rochester, New York

John H. Gibbons
U.S. Congress
Office of Technology Assessment
Washington, D.C.

William G. Gillespie
US Department of Energy
Washington, D.C.

Robert J. Glasser
LeBouf, Lamb, Leiby & MacRae
New York, New York

Jack Goldman
Aluminum Association, Inc.
Washington, D.C.

Ruth Gonze
American Public Power Association
Washington, D.C.

William P. Gulledge
Chemical Manufacturers Association
Washington, D.C.

Charles Hakkarinen
Electric Power Research Institute
Palo Alto, California

Jeremy M. Hales
Battelle- Pacific NW
Richland, Washington

John D. Hall, Jr.
State Corporation Commission
Richmond, Virginia

D. Alan Hansen
Electric Power Research Institute
Palo Alto, California

Janet Hathaway
National Wildlife Federation
Washington, D.C.

David Hawkins
Natural Resources Defense Council
Washington, D.C.

Nelson E. Hay
American Gas Association
Arlington, Virginia

Ned Helme
Alliance for Acid Rain Control
Washington, D.C.

Rosemary Henderson
Division of Energy Regulations
State of Virginia
Richmond, Virginia

George Hendrey, Ph.D.
Brookhaven National Laboratory
Upton, New York

Eric Hennen
Dairyland Power Cooperative
LaCrosse, Wisconsin

Donna Hickman
Resource Recovery Magazine
Washington, D.C.

Dr. George M. Hidy
Desert Research Institute
Reno, Nevada

Jean Huff
University of Maryland
College Park, Maryland

Adele M. Hurley
Canadian Coalition on Acid Rain
Toronto, Ontario Canada

Marty W. Irwin
Hoosiers/Economic
Development Commission
Indianapolis, Indiana

Richard P. Janoso
Pennsylvania Power & Light
Allentown, Pennsylvania

John J. Jansen
Southern Company Services
Birmingham, Alabama

Tahane Joyal
National Wildlife Federation
Washington, D.C.

Anthony R. Katz
West Associates
New York, New York

William R. Kelly
National Bureau of Standards
Gaithersburg, Maryland

Dr. Edwin H. Ketchledge
Adirondack Mountain Club
Bloomingdale, New York

Nancy Kete
Johns Hopkins University
Baltimore, Maryland

Kenneth D. Kimball
Appalachian Mountain Club
Gorham, New Hampshire

Elise Kirban
US Environmental Protection Agency
Washington, D.C.

Barry R. Korb
Environmental Protection Agency
Rockville, Maryland

John R. Kruse
Environmental Research & Technology
Andover, Massachusetts

Wrandy Kubetin
Bureau of National Affairs
Washington, D.C.

J. Laurence Kulp
National Acid Precipitation
Assessment Program
Washington, D.C.

Thomas Kusterer
Department of Environmental Health
Baltimore, Maryland

Philip Lapat
Brookfield Center, Connecticut

Chen Le-Tian
National Bureau of Standards
Gaithersburg, Maryland

Dr. Miriam Lav-An
Environmental Monitor Service
Camarillo, California

Cynthia Lecompte
University of Maryland
College Park, Maryland

Gene E. Likens
Institute of Ecosystems Studies
Millbrook, New York

Perry Lindstrom
American Petroleum Institute
Washington, D.C.

Lori Lopa
Center for Environmental Information
Rochester, New York

Glenn H. Lovin
Resource Recovery Magazine
Washington, D.C.

Alan A. Lucier
National Council for Air and
Stream Improvement
New York, New York

Maris Lusis
Ontario Ministry of Environment
Toronto, Ontario Canada

Lester Machta
National Oceanic and
Atmospheric Administration
Silver Springs, Maryland

James J. MacKenzie
World Resources Institute
Silver Springs, Maryland

Mike Malloy
Coal and Synfuels Technology
Arlington, Virginia

C. Robert Manor
Potomac Electric Power Company
Washington, D.C.

Michael A. McCord
Department of Justice
Springfield, Virginia

Frank J. McDowell
Indiana Environmental
Public Service
Plainfield, Indiana

Robert McFadden
Motor Vehicle Manufacturers
Association
Washington, D.C.

Kelton McKinley
General Research Corporation
McLean, Virginia

Thomas R. Monroe
NYS Department of Environmental Conservation
Ray Brook, New York

Chris Neme
Center for Acid Rain and
Clean Air Policy Analysis
Washington, D.C.

Brand Nieman
US Environmental Protection Agency
Washington, D.C.

Calvin M. Ogburn
Carolina Power and Light
New Hill, North Carolina

Dr. John Ondov
University of Maryland
College Park, Maryland

Michael Oppenheimer
Environmental Defense
Fund, Inc.
New York, New York

Neil Orloff
Center for Environmental
Research; Cornell University
Ithaca, New York

Donald H. Pack
Certified Consulting Meteorologist
McLean, Virginia

Eustice Parnelle
Florida Power and Light
St. Petersburg, Florida

Dr. Ralph Perhac
Electric Power Research Institute
Palo Alto, California

Jean Perih
Office of Congressman Don Ritter
Washington, D.C.

Ranard J. Pickering
United States Geological Survey
Reston, Virginia

Pam Prah
Coal Week Magazine
Washington, D.C.

Mary Pratt
Center for Environmental Information
Rochester, New York

Laurence Pringle
Nyack, New York

Theresa Pugh
National Association of Manufacturers
Washington, D.C.

Jefferey Quyle
Indiana Department of Commerce
Indianapolis, Indiana

Gary Randorf
Adirondack Council
Elizabethtown, New York

James L. Regens
Natural Resources;
University of Georgia
Athens, Georgia

John J. Reilly
Pennsylvania Electric Company
Johnstown, Pennsylvania

David S. Renne
Battelle-Pacific NW
Richland, Washington

Sue Roussopoulos
USDA Forest Service
Washington, D.C.

Sandy Royster
Commonwealth Edison
Glen Ellyn, Illinois

Scott Rubin
Office of Consumer Advocate
Harrisburg, Pennsylvania

Dan Salkovitz
Air Pollution Control
Richmond, Virginia

Perry J. Samson
University of Michigan
Ann Arbor, Michigan

Greg Shieffer
Life Systems, Inc.
Cleveland, Ohio

George Schroeder
Limestone Products Corporation
Prompton, Pennsylvania

Paul Schwengels
US Environmental Protection Agency
Washington, D.C.

David J. Shaw
Department of Environmental Conservation
Albany, New York

Deborah Sheiman
National Resources Defense Council, Inc.
Washington, D.C.

Carol Simmons
Colorado State University
Ft. Collins, Colorado

Dr. Nancy Paige Smith
St. Mary's College
St. Mary's, Maryland

Bonnie J. Stoner
SURRES
Washington, D.C.

Harry L. Storey
Alliance for Clean Energy
Denver, Colorado

Frederick W. Stoss
Acid Rain Information Clearinghouse
Rochester, New York

Dr. David G. Streets
Argonne National Laboratory
Argonne, Illinois

Ellis E. Sykes
Albany, New York

Michael L. Teague
Hunton and Williams
Richmond, Virginia

Ralph Tedesco
NYS Electric & Gas
Binghamton, New York

Jean Thompson
Center for Environmental Information
Rochester, New York

Elizabeth Thorndike
Center for Environmental Information
Rochester, New York

George Tomlinson
DOMTAR, Inc.
Perrot, Quebec Canada

Gurdal Tuncel
University of Maryland
College Park, Maryland

Tom Ulasewicz
Adirondack Park Agency
Ray Brook, New York

Janice Wagner
Alliance Technologies Corporation
Bedford, Massachusetts

Nancy E. Wagner
Georgia-Pacific Corporation
Washington, D.C.

William Wagner
Center for Environmental Information
Rochester, New York

Linda Wall
Center for Environmental Information
Rochester, New York

Dr. Amy Walton
Jet Propulsion Laboratory
Pasadena, California

Dr. Walter Warnick
US Department of Energy
Washington, D.C.

Neva Welch
Department of Water & Natural Resources
Pierre, South Dakota

Marchant Wentworth
Izzak Walton League
Arlington, Virginia

James C. White
Center for Environmental Information
Rochester, New York

Carol Willey
Center for Environmental Information
Rochester, New York

David R. Wooley
Environmental Protection Bureau
Office of Attorney General
Albany, New York

James W.S. Young
Atmospheric Environment Service
Downsview, Ontario Canada